About 'Spear

A minute before I met the AUTHOR OF THE UNIVERSE, I was wandering the world quite comfortably attired in a pleasant naïveté. I didn't know where the Universe came from and I had no formula for bringing about a better world that didn't include a giant asteroid — though I had been seeking such a formula for millennia. But, all I wanted just then was to wangle a nourishing lunch via the application of artful storytelling. Could I have guessed that the crusty octogenarian in the pub was the AOU?

To save you some trouble, may I offer that any entity introducing itself as the Author of the Universe should be avoided? This applies to the brightest of you too. And to reinforce the point, Melissa, a former Princeton graduate student, now the smartest person on Earth is also an acquaintance of the Author of the Universe. Even a skeptic in training has learned the hard lesson of undiscriminated acquaintance. On the plus side, she no longer falls asleep in class. She no longer has to go to class.

So, because I was trapped by the ethics that require one to reciprocate value when receiving a free lunch, and because Melissa feared being bored to death, we joined together in a moderately challenging endeavor — to facilitate the survival of our species — the history of which can be found in this portion of the Force Bottle Journals. May it serve your purposes now. — *TM*

Editor's Note:

Thodkin's Spear is an immensely readable — and highly original — "fictional" grand tour through the big issues facing mankind at this evolutionary turning point: longevity, morality, truth, evolution, reason, religion, an afterlife, and the advance of science and technology, just to name a few. Scott Ellison paints possibilities with a wise and colorful brush in this literate, entertaining and important book.

Okay, we're biased, but that's what we think about 'Spear.

Praise?

"Extremely deep, philosophical and whimsical simultaneously.
Look out: a new writing style has topped the horizon. S. Ellison emerges a modern James Joyce-meets-Douglas Adams: unique insight, humor, groundedness, refreshing imagination and thought provoking perspective..

— K. Palmer, Rimrockin' Times Herald (It's what it would say if it it existed. — K.P.)

SELF-NOMINATED for
BEST-BY-FAMILY-MEMBER AWARD

"...unmitigated work of...Whoowee boy - you just watch: you will have groupies.
Having said that, I will also say that the actual storyline is a thinly veiled vehicle for relaying richly developed and interestingly archetyped discursions. Robert M. Persig's chataquas [Zen and the Art of Motorcycle Maintenance] have nothing on you! There is nutritious food for thought between the lines here."

— Bob Daniels, Technology Guru & Unix-Geek

"I recently discovered a box of 10,000 self-powered lightbulbs...
Hiding inside the covers of Thodkin's Spear. No, it's not a light tale, just one of luminous purpose: painless therapy for stiffened minds. Deliciously eloquent, irreverent, humorous, inventive, subtle and wise, it's also so original that Originality herself will want to sign up for night classes with the author."

— E. T. Ellison,

Author of The Luck of Madonna 13
(2002 Book of the Year winner)
(He is Scott's brother...so?)

THODKIN'S SPEAR

Force Bottle Journals

Entry No. 111, Book 1

by Scott Ellison

QUIRKHAMMER PRESS
LOS ANGELES & VIRTUALLY EVERYWHERE

COPYRIGHT

This is a work of fiction. Names characters, places and incidents are either the product of the author's dementia or are used fictitiously. Any resemblance to actual events, locales, organizations, or persons, living or dead, is entirely coincidental and beyond intent of either the author or the publisher.

Publisher's Cataloging-in-Publication data (provided by Quality Books, Inc.)

Ellison, Scott

Thodkin's Spear / by Scott Ellison -- 1st ed.

p. cm. --(Thodkin's Spear)

ISBN 978-0989144834

1. Science fiction. 2. Fantasy fiction 3. Social commentary I. Title.

QuirkHammer Press - books for thought™
an imprint of QuirkHammer LLC

13636-109 Ventura Blvd
Sherman Oaks, CA 91423 USA

epress@quirkhammerpress.com
www.quirkhammerpress.com

Dust jacket and page xi illustrations by Scott Ellison,
Frontispiece illustration by Scott Ellison
Endpaper design "Force Bottle Version III" by S. Ellison, www.thodkin.com

QuirkHammer Press

FIRST PRINTING, JUNE 2013

DEDICATIONS

TO JB

- FOR UNEARNED TOLERANCE -

... AND A SPECIAL INDICTMENT OF ETE WHOSE ENCOURAGEMENT AND BLIND BROTHERLY LOVE CAN BE DIRECTLY BLAMED FOR WHATEVER IS OR ISN'T TO BE FOUND WITHIN THESE PAGES.

... AND TO GW FOR GIVING ME SOMETHING TO TALK ABOUT.

PREFACE

It could be said that there is no evidence of the existence of a 10,000-year-old man, nor a visitation of any being capable of authoring the universe. Since it is not to their advantage to be known, you are forced to accept their existence absent physical evidence. You may do that if you like, needlessly risking the label of fool. But even more seriously, it is fervently hoped that if you are reading this work some time in the distant future, out of the context of the time of its writing, you do not attribute deity to the characters of this story. They are, for goodness' sake, part of a work of fiction. Whether real or fabricated, they are just people, or not, as the case may be.

Another prefacing interjection concerns hidden subsocieties intent on improving our lot, no strings attached. Critics may question suppositions and intimations regarding the existence of organizations interested in forwarding the species. It is true that one must look well away from the bright lights focused on the popular actors and puppeteers manipulating most of contemporary society in order to discern the activists striving out of limelight. But, one day, perhaps soon, or possibly later, you will surely find yourself part of a dynamic neo-humanity. A humanity that has broken the shackles of superstition and leaped toward comprehension of the cores of wonder.

A Mundane Note on the Contents

There are no chapters to this story. It is a story that just goes front to back, told as it happened and as it will continue to happen. True, because of the uninterrupted flow, there is a small downside for those who feel folding the corners of a page to mark a reentry point is an abomination in the sequence of all things. To help the anti-abominists, or non-abominists, I have included sequence markers. I admit they are only approximations of chapters, but there are no chapters in life. We don't get to stop, take our rest, wake up and commence a new chapter, unless one is an amnesiac, in which case one might have to start the story over at the beginning. But, if you wish to use them, I have marked the geographical locations of the story. If you remember where things were happening when you dozed off, you have a chance of getting a night's rest confident in your freedom from self-disgust. For speed readers, enjoy your luxurious continuity.

A note from the Archive CI:

I, the Archive CI (Created Intelligence), am responsible for the formatting of this book. I have opted for a brief explanation to aid unenhanced minds that may undertake a reading.

Thodkin/Melissa dialog and musings are in this text format.

Metaphorical Stories are in this text.

Thodkin Quotes

Thodkin Reference Name

Thodkin Stories

Recipe Subheader

Recipe Instruction Ingredients

Recipe Instruction Line

Recipe Instruction child line

Notes in body of text

CI Archivist narration support

Lists in Body

List Point in Body

Sidebar with interesting information, or not

Sidebar points

footnotes

It is hoped this is will assist deciphering an otherwise most convoluted and weakly structured intertwining of dialog, context and meaning.

CI_132 Level 0, Archivist Guild Rivendell II

Contents

THODKIN'S SPEAR

Force Bottle Journals

Entry No. 111, Book 1

by Scott Ellison

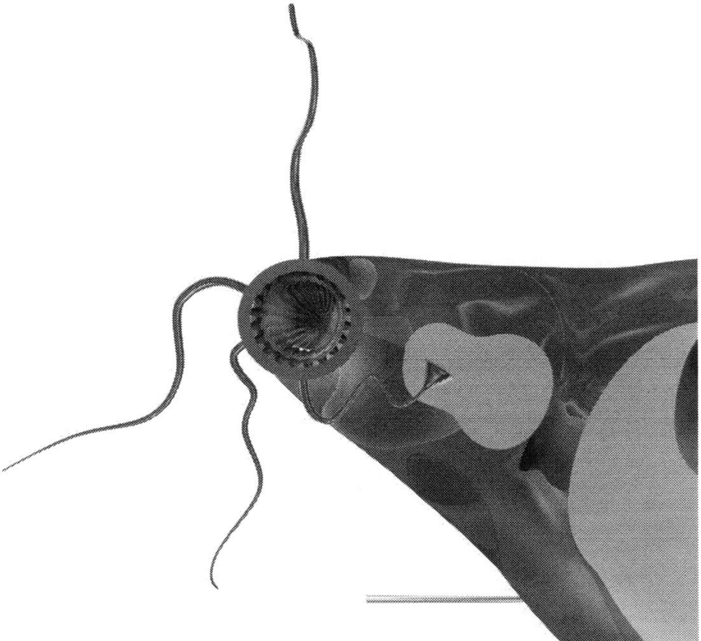

Grass

With the strap of her red-tipped cane looped around one sleeve, the pharmacist considers the fringe of chaos. Except for an idle, nerveless tapping, she stands unnaturally still. She is a tall, thin and ominous shadow, wrapped in a hooded floor-length cloak. She is cloaked and edgeless; and can only be discerned as a wavering intimation of deep gray apparel — with just the one little tap-tap-tap twitch, skulking away from her supernatural attention.

Or perhaps she wears a white lab coat, her hands in pockets, pondering.

Or other times no coat at all and no pockets, just a light summer dress cut from cloth of cheerful floral print.

Or possibly she wears nothing at all and dances about the apothecary shop in abstract, vivacious innocence. Maybe she even dances, as such, into the fog of chaos itself, hoping for acquaintance with an exotic partner ready to conga away from the asynchronic rhythms of the fringe.

However she is appointed, she is irrefutably at work exercising her artful skills, one foot in the fringe, the other in the quaint *laboratory of irreverent concoction*, known also as *the place of first contact*. That is the place where identity emerges from the noisy fog of chaos. It is where she gently coaxes innovation into shapely crystal bottles only then to absently stock them onto the cluttered shelves of reality; but not before, in chuckling spirit of full disclosure, they are labeled *caveat exspirare e' dominiarium fidere.**

She is the anarchist spirit of any place where complexity dangles bau-

• • • • • • •

*Archaic Latin: Literally translated: "Beware expiry" and "look out for tyrannical deception or beware, no promises". In contemporary vernacular: "Warning, unknown shelf life," and "Caution, sold 'As Is.'"

bles sparkling with rays of fetish or thrusts cold obelisks of enrapturing obsession before the cognitive mind's eye...or paints itself on the back of evolution's turtle to take chances with mutation.

Facts suggest, or perhaps only intimate, that she was wearing her dark cloak the day her sensitive fingers penetrated the minds (sic) of the Author of the Universe. But these facts arrange the order of probabilities. And from the resulting quantum, cosmic bookkeepers yielded odds leaning across the veil of certainty that it was the Pharmacist who, from the networks of virtual neurons sewn through whole universes, shaped the idea that its "beingness" — that is, the entity known as the Author of the Universe — required a brief consultation with we lowly humans.

But, and this is not conjecture, she was surely dancing in an avant-garde mist, hands clapping seductively from side to side, her hips taunting and beckoning the mischievous sprites of perverse relation, as I continued my long walk that morning of eventual introduction. It was to be one more day, of many, that I had been designated the whipping boy to feel the lash of her flagellant medicine. It was a day that, in spite of my abilities to resist dangerous entreaty, the apothecary would have her way with me. It is just luck I was in the mood for looming truth. Such is not the case every morning — irrelevant, of course, very few control the loom of truth.

As a small aside — in case these questions rose to your attention — no, I cannot disclose the relevance of the blind pharmacist's fashion, nor its observational prominence. Like so many things, these just are what they are until they are something else.

Now I'll explain grass.

Disclaimer: Rhetoric Motivates Itself

If you have ever walked through tall grasses on a dew-covered field as the sun rises, you already have an understanding of the nature of life. That is, the true nature of life which goes by many descriptions but reduces down to just one pure statement of truth.

No, that is not true.

Well, let's allow for more than one truth. What if there is more than one pure statement of truth? What if there is the one I know, and let's say you have one, too. So, there are two absolutely true natures of life. Or maybe there are three because there is one we haven't heard about.*

Main Entry: **pru·dence**
Pronunciation: **'prü-d&n(t)s**
Function: *noun*
Etymology: Middle English, from Middle French, from Latin *prudentia,* alteration of *providentia* - more at **PROVIDENCE**
Date: 14th century
1 : the ability to govern and discipline oneself by the use of reason
2 : sagacity or shrewdness in the management of affairs
3 : skill and good judgment in the use of resources
4 : caution or circumspection as to danger or risk

Okay, however many there are, each has one very important thing in common; each one... is a prescription filled by the blind pharmacist. Therefore, not only is it questionable that the contents born of flowing evolutionary chaos will be efficacious — that you can live by them — but the direction on the label may be difficult to decipher, written as it is, without the benefit of illumination. It is common to be offered the alluring drug of truth, and something else to find illumination.

Again

Life is seeing the sun rise on a clear cool morning while walking through tall stalks of country grass bent with morning dew. This is probably not true pervasively; it may only be a component of life, but it allows me to say that to feel the beauty and wonder of life's true nature you will have to get your feet wet and move forward into the do.

••••••••

* The cultural historian William Irwin Thompson has approached these issues from a quite different direction and has articulated most intriguing possibilities. For him: "Values are not objects, they are relationships. When you overlay one pattern with another, a third pattern emerges, a moiré pattern" (Thompson, p. 38). He argues that: "Truth cannot be expressed in an ideology, for Truth is that which overlights the conflict of opposed ideologies....The Truth cannot be known in an ideology, but it can be embodied in an ecology; anything less does violence to human nature and to human culture" (p. 36).
Commentaries from *Encyclopedia of World Problems and Human Potential* "http://www.uia.be/values/valcom_bodies.php?kap=16"

Anyway, this crisp morning I had been walking toward Princeton across a dew-covered field since sun-rise, seeking truth, when after several hours I abruptly noticed that in all my years I had been unable to reduce life down to one pure statement of truth — *the General Theory of the Truth of Life* that could be used to relate all the questions that perplexed me into a single illuminating Gestalt providing *the correct guidance* for achievement — and then I realized that my pant legs were steaming. This steamulated me to think that it would be nice to dry out and get something to eat. Maybe the truth of life could not be found in a field of tall grass at sunrise. I hadn't found it there, so far. Maybe the truth could be found in a nice warm eatery surrounded by nutrition. Fairly certain the true nature of life was playing hide and seek with this particular field and the field was It, still eyes closed and counting, I resolved to seek truth in another place it might reside… near lunch. So, if truth isn't to be found in your current location or circumstance, and if you don't know, with certainty, where it is likely to be found, consider it a good idea, as good as any, to seek it in a nice warm, dry place filled with matured possibilities for a pleasant meal. So much is simple prudence.

The truth is that lunch is the reason for this story. The truth is that not finding the true nature of life in a field of grass is the reason for this story. That is two truths, both *the* reason, because I thought of them sequentially as I was writing. If I had made notes while conducting a survey of people who knew truth to be something other than lunch and, not finding truth in the grass, then compiled a list of truths, each of my two truths would have just been *a* truth, which says something about living in the present; sometimes a thing is given too big a slice of pie. Still, ignoring commentary about living in the present and existing in sequences, neither truth can be considered *the big, sought after by all,* true nature of life — which might be found in the results of a survey, or may someday be found or not, probably not. Truth can be elusive.

It is a small lesson that you might be occasionally surprised at what you will have to swallow because of a free lunch — which is fine with me. I long ago resolved to pay for my free lunches; and to watch out for hidden costs in a free lunch, seemingly an illogical statement at the core, but I am now speaking of the a*lleged free lunch*; this one cost me dearly, though I do not say I regret it. And the cost was not in the coin of insightful stories that I am accustomed to proffering. So it is perhaps a durable life truth, though not the only one, that sometimes free lunches are not free.

There may be more to be digested about lunch in this story — that remains to be seen. I have never told the story before, so even I don't know what is in it. I am — for the moment — living in the present.

The Known and Responsibility

What I do know is, that because of my luncheon's benefactor, I have adopted the responsibility to put forward predictions about what is, in the short term, going to be the nature of modern life.

It may be considered that I am prophesying. But, I would protest being called Prophet. Nor am I in any way a fate-ist. Forces of the universe will interact or avoid interaction in a much too complex lambada to predict with accuracy which vital torsos will entwine. Any of my predictions are no more than trend analysis, or masked hard information disclosed by insiders — both counter, in method, to prophetic expostulations. Prophets do not predict with Cartesian ordinality of time and space. This is a key to their success. Prophesies can't be wrong if they don't say where and when.

Poetry?

In the order of life
In the time between lunches
In the pondering time
Is the knowledge of spaces
Two come to mind
There's the space of the problem
Always found easy
And the space of solution
Unsuccumbing entreaty

—Anon (for good reason)

And fate-ists or destinists swim in placid currents of impotent obsolescence. Genghis Khan professed destiny apparently unaware of choosing second fiddle over first chair. How long did Kahnism last? About a minute after his death, as I recall. "The Khan is dead, long live the Khan — Okay, Jesus I'm starved! What's for lunch?" Clearly, destiny was trumped by divinity.

Then, if you add divinity to prediction, you get prophecy. Divinity trumps all power levels, hence Divine Right and the lasting ideaisms of the prophets. Prophecy conjoined with deity is the strongest, most versatile conservative Möbius thought trap so far invented.

Me Thinks Ye Prophetess Too Much

But as powerful as clever prophets have been shown to be, I shrink from the title. Competition is fierce and the game is old. Contemporary, nouveau-prophets are playing catch-up to those more established. I surely would not want to be a come-lately prophet. Besides, it is just a matter of time before you get included in a holy war — inevitable if you include divinity in your projections. Sometimes more than once and for extended durations you will be asked to defend your unprovable truth against the stubborn holder of another, even risking mortality — "My deity can whup your deity." Good, but how will you tell it has? And risking life in dispute over the unprovable? Life at least can be observed by the senses — or mis-observed and non-sensed — as I suppose, within certain boundaries, is one's inalienable right. Actually, the advent of religious tolerance as a member of socio-political correctness is a direct descendant of religious pacifism. Religious tolerance was invented by a style of pacifist, or more specifically, people unwilling to lose their lives over the cosmically unsubstantiable. Needless to say, knowing what I know, that all seems beside the point, because I claim no ownership of knowledge passed by a deity. The guy in the pub, to his credit, did not claim the role. So, let me sum up this beginning by suggesting that life is complex and propose that oversimplifying it only diminishes our abilities and effectiveness.

There May Be More than One True Nature of Life...or...

Or all natures of life might better be considered *trueish*. It might serve us to adjust the natures we perceive up and down on a scale of fuzzy truth, *trueishness*. And furthermore, being right about a prediction of the future does not make one a prophet, unless they actually predict a god is going to arrive in Princeton on a specified date within the prophet's own lifetime, and it does. Of course, one event is not a trend. But, that prophet would get my vote for most-true prophet, highest *trueishness* of prophecy to date. And I seem to have gotten sidetracked into discussing divinity/destiny/space-time-ordinal incongruities and conflicts, when I just wanted to avoid being mis-termed a prophet when it is discovered, because of the persistence of this little book, that I have been, on the whole, irrefutably right about the future and the need to prepare for certain characteristics of coming existence caused by the rapidly approaching technologies driving cognitive-evolution. Just so.

Let's Do Lunch

A high order truth for me was that a pondering morning had slipped past without enlightenment from the cosmic nova of truth and I needed a bit of lunch.

Rhetoric Gives Way...Action

So, I pushed open the traditional heavy wooden door and stepped out of the weather (it rains in Princeton). It took a moment to adjust to the room light, but while I was standing there I noticed a man in a corner booth farthest from the men's room. I figured he was brighter than most. Given a choice, bright people sit as far away from the men's room as they can…at least in a pub they do. My second criterion, which relates to benefactor quotient and inebriation, was also fulfilled (or so I thought at the time). The table was strewn with empty beer glasses, all on his side. I say strewn, but they were oddly patterned in a little beer glass spiral ending with the half full glass in his spindly fist. From the tight, lip to lip spacing of the design, I guessed he was planning on being at the booth two or three more days. Maybe I could interrupt.

He looked up as I evaluated the possibilities for lunch — good chance of a burger and fries. He looked like a burger and fries to me. Then as I considered the potentials of slightly devious tactics in an overall consumptive strategy, he just waved me over. At my arrival he said, "Hey, young fella, can you watch my beer… gotta go to the porcelain two-holer they got back there. I'll spring for a burger and fries if you keep these kids from rustling my booth..

Hmmm, I thought, this is somewhat odd. "Sure, leave it to me. And I'll take you up on the food offer." He hobbled away, his rickets-ridden gait about a six inch shuffle. I understood why he needed a booth sitter. I thought the guy should get more sun. Turns out I was somewhat wrong in my diagnosis.

The waiter came by, asked me what I wanted. I said "Burger and fries." He said, "Excellent choice." Just so, and that was that.

I have learned that a good burger is a joy forever... Okay, a half hour. And one thing that adds spice to that joyful way is a *burger gespendet worden.** I was savoring the priceless grease drippings on my plate by smooshing some loose fries in the fragrant greasy puddles and then popping the spud in my mouth from three or four inches out, sort of a three-point play of fries devourment, if you understand the scaling reference from roundball, when my booth partner and benefactor returned from his trip.

He jumped right in and said, "Can't beat a free burger. Makes fries smoosh just that much better...good shot! Pleasure to make your acquaintance. I'm the author of the universe." This was delivered as a straight line punctuated subtly with a twinkle in his eyes.

Few people surprise me these days. When you have lived several thousand years, when your life goes back to early recorded history, when in fact you are the inventor of recording itself, it could just be said without a boast that you have learned a thing or two. And one thing I have learned...people who are surprising have spent serious effort becoming surprising and they definitely have tricks up their sleeve. The best use both sleeves.

On with It

"Call me Thodkin, Thodkin Marblemeister...you are the author of the universe? That seems unlikely to me. I have met several over the years, all of whom have turned out to be frauds or deluded. How can I prove your bon af ides? And shouldn't author of the universe have initial caps?"

•••••••

* Donated hamburger.

"Well, you could just believe me. Thodkin is an unusual name.

"It is one less common. As for just believing, yes, I've heard that one already; its status is 'classic.' Unfortunately, I feel it necessary to see a manifestation of physical proof, or at least precise logic and reason, immutable even through the ambiguous language of the current times. I wish to see evidence of something other than my own wishes. I wish wishless observability both measurable and sensory specific. I seek to see, feel meaningful fluctuations in air pressure and if it won't kill me, tactility."

"I see your difficulty. Not nice calling me a liar though. There are systems in the Void where you might suffer a showdown. Aw, don't worry it. Could be, if I tell you some of the story of the universe and explain my presence in this little college town, indeed this backwater world, you will experience the truth. Maybe a cowboy can know the truth without steppin' in physical proof. Perhaps the truth can be experienced by noticing the dramatic tango of order and chaos in complexity, or a fancy two-step." The author allowed me an interjection before he continued.

"Perhaps I will be introduced to the Blind Pharmacist herself, and together we will spawn a brood of comprehension. Or perhaps she will combine dusts, compress them into a tablet of epiphany, or a capsule, or an invigorating syrup." I filled available air space and he followed with a brief reflection and temptation.

"Similar fickle liaisons are reality in every universe. Would you like to attempt one? I've got time. I'm working on a problem and the schedule of calculation is of uncertain duration."

It may be different with you, but I like to pay for my food. I don't mind hand-me-downs of clothes or other necessities, but I feel a consumable should be paid for. It seemed to me, in my naïveté, that just this once I could listen to a story. Normally, the burden is on me; you may know me as Thodkin the Storyteller? I thought it might be a nice change but I was hoping the beer glass spiral wasn't evidence of the problem he was working on. "Sure, complexity is what all things evolve towards. I can imagine that an explanation of the universe might involve some level of complexity, and discussion of it."

He paused, although I know he didn't really need to think about what I had said, then kind of grumbled. "This horse-of-a-different-feather does anyway. It wasn't what I had in mind. But, I'm not too old to learn a thing when it slaps me in the face."

Who is?

If someday you sit across from an old guy who says he is the author of the universe, speaks in a hodgepodge of archaic syntax from western cinema intertwined with expressions of near Oxford perfection, may I suggest that you seek seating elsewhere, by the men's room if necessary, and forgo lunch if need be, unless you have no love for your perception of the universe. I'd not been given such advice, so I naively listened attentively, fulfilling my end of the bargain and now am burdened, a victim of fine print. Oh well, I really had nothing more pressing.

And so it went.

Hook, Sinker and Consequences of Not Reading the Fine Print

He began slowly. "I ask you what your most precious possession is."

I have tried to be unencumbered by the superfluous. Live 10,000 years and see how many things last. I just got tired of seeing my prize acquisitions fall apart, disintegrate, be stolen or worst of all turn out to be a fad. So, I began to require everything to have deeper purpose. Gradually, that reduced ownership of artifacts to the clothes on my back. Not that these clothes had deep meaning, but they were, metaphorically, my shell. I am a 10,000-year-old tortoise. When I become too conspicuous, I change my clothing style. When I restyle my clothes it is like pulling my head back inside my shell. Unlike a makeover that reveals something of a person that has been hidden, people don't see me, just the clothes — and I am hidden. I have found that to be very useful. I planned, now that I was in a college town, to find the nearest Salvation Army store and pick out a "professor" outfit. An ensemble of Levi's, runners (tennis shoes, sneakers) and a classic brown tweed sport coat. It has also been known as the Hollywood Director's Formal Wear. Redford, a medium-sized icon of the time, looked particularly comfortable in this attire. And, after careful leafing of clothing racks it would be the perfect college town getup, fetching and quite affordable. From the moment I would leave the store, people would want to know what course I taught. They would listen to my stories and take notes. They would add meaning to the story that was never intended. They would find ways to validate my stories with their own lives. And all because of the way I had painted my shell. Though they are making progress, it will be some time yet before humans perceive beneath the surface first. It will be a while before "the clothes make the storyteller" has lost validity.

So, maybe it was my shell that was most precious to me. But really, making the effective use of my clothes was just one of the things I had learned in my travels. It was precious, but it wasn't my most important possession. The clothes don't really make the man, as this story should prove conclusively. They are just a tool for expressing commonality or distinction. You know, you look like me, you must know something about me, you must be like me. I like people who are something like me. No, clothes are just something used to communicate. They have long since ceased importance as protection from the elements. Now, they are assertions of self and community. Important though it is, my shell isn't my most important possession.

"I don't know. I don't have much."

He looked at me and said. "For most people the question is bitin' off more'n they can chew, but they answer it anyway. Like a dude drinkin' downstream from the herd, they spit a mouthful of cliché and strong emotion. Well, I also pack an answer to the question. It is an answer that has cost me plenty to learn. And it is an answer that lightened your purse, too.

I translated his statement as best I could, then I told him it seemed like a simple question, but after realizing I had become such a minimalist, it had taken on a surprising complexity.

"How much do you know about complexity?" he asked.

"It is like art. I don't know much about art, but I like what I know. And I think I have simplified my life so that I can better understand the complex. I have begun to theorize that I can learn to sense the nuance or shifts of complexity. It requires a rich and accurate internal model of the universe, an achievement I can only be unsure of. It is not enough to know a narrow domain very well, to specialize. Rather, I'm thinking it requires comprehension of relationship…pan-domain. The theory is more of an intuition at the moment."

"Welcome," he said. "You are entering the next stage of human development, unless I miss my guess. And my guesses are not usually guesses. In fact, part of the information I am waiting for will validate this assertion. Humans are nearing an avalanche that is the direct result of their increasing ability to discover patterns in the apparent noise of complexity."

I asked him if he was pleased. After all, being the AOU, seeing his creation move to a higher plane, must be rewarding.

He laughed. "Are you kidding? This isn't what I wanted. Anyway, it's personality."

I told him I hadn't quite understood that last comment, he said. "Personality is your most important possession."

What about intelligence? What about good looks, or great DNA or a rich family? What about nice weather and plenty to eat? How about loyal friends and good lovers or loyal lovers and good friends? Maybe we should think about this carefully? Personality? What about life itself?

He seemed to read my mind a bit and countered by saying, "Intelligence is nothing without personality, neither is life. It is a car with no steering, no brakes, no clutch for shifting gears, no inspiration or direction. It is a brick alone, distinguished from architecture; an artifact. Apply personality and intelligence becomes a building block. Well, emotion provides the inspiration, and unknown to most, intelligence breathes stamina into emotion by focusing it, but personality is the tool that carves shape from the block of emotion your evolution has stuck you with. Without a cognitive personality capable of self-referential evolution, you ride the horse in the direction it's going; personality lets you get on the horse and ride off in all directions."

Some may consider the last item something to be avoided. I asked him why he designed emotion to be so brutish.

"Again, you lengthen my extremity. I didn't have anything like emotion in mind. It's a tick in my hide. Well, actually, it is a bug resulting from a bug but who will even notice? See, I never intended there to be Evolution. My plan was that DNA would produce a nice straight protein. With these very predictable proteins I had in mind I'd get a very placid organism. I was looking for worshipers, puppets. 'Yas Boss, whatever you want Boss.' Look what I got. I was building a vacation resort. I wanted a puzzle with all the pieces the same size and shape. And I got a chaotic evolving soup of ignorant know-it-alls."

I said, "'Vacation resort' is a puzzling metaphor for a world of life..

He said, "That was no metaphor, that is what I was building. I wanted

• • • • • • •

* Ours is a lab model Universe Containment Sphere with Force Backup guaranteed for 10 Standard Universal Units or approximately 37.9 billion years Experiment Reality Time (ERT), OMB (or money back). The Stealth feature resolves earlier problems with a mysterious wallpapering effect that appeared on the inside of the bottle and was noticeable with superior instrumentation as a flower pattern; we would know them as daisies. The pattern was really just a fractal, but most disturbing to the scientist who first discovered it. The Stealth feature makes it appear as though the contained universe can expand to infinity. In practice, any matter or energy reaching the edge of the containment barrier is reduced to Force Bottle Quanta and shunted back to the Quantum Source hidden artfully in the base that keeps the Force Bottle from rolling off of the Lab Bench. A recent fad of the flambicious is to have active universes on one's livingroom coffee table or awards shelf. Small ones have been seen on necklaces at some chi-chi parties, although this is somewhat risky. If the Quantum Source fails, party goers receive an unpleasant surpise, or if payments haven't been kept current, the necklace quickly forms a localized black hole binding anyone nearby into the creditors' flux source until payment is arranged. You laugh, but this has happened.

servants, not intellectual competitors."

I asked him why he didn't just start over.

"I did, on this and many other planets. I smashed a huge rock into this planet. It was obvious things were out of hand. I was hoping for a restart, but when my back was turned you spring up. I'd have done better with the Tyrannosaurus rex. Sooner or later there would have been one really engorged one, and I could have built a nice lodge somewhere it couldn't reach. We have rules…now I can't shut down the Force Bottle* until all the energy in it is neutralized. That's going to be another trillion trillion trillion years. I'm just writing it off. I'm here now to analyze where I went wrong. Then I buy another bottle and start over. Wish you luck."

"What is a Force Bottle?"

"This universe as you know it is actually contained in a Force Bottle Version III with a Stealth upgrade. It is my first one, best I could buy. We don't have land quite like you know it. We exist in little self-sustaining matter pockets in what might not be the nothingness. If we need a break, or a vista, we make one. I'm going to have a nice place where I can sit quietly under the stars in a toad floater and stare at nothing."

"So are you gods?" I pondered the idea of gods infatuated with cowboy slang, also unable to construct a coherent aphorism.

"As far as we can tell, there are no gods, just entities with more knowledge. It's like magic until you can do it yourself. You have probably heard that before..

"Yes, it is a mantra. So, where does nothingness come from?"

"Ya got my dogie tied…might be magic. Or, alternatively, we must accept a different starting point of analysis. For example, it is possible to conjecture that the universe was always there, and nothingness includes it — this requires uninventing the intuitive definition of nothingness — and seems to dictate that nothingness is timeless, therefore always there; more accurately — nothingness is.

"Or there is just no such thing as nothingness. If one employs the weak anthropic principle, then if nothingness exists, it must allow for the observer. I'm beginning to conjecture that nothingness can never have existed because its definition excludes us — the observers. True, there are theories that an infinite nothing will have some probability of spontaneously creating a something, but that then becomes a certainty. Therefore, the universe always existed — at least probably. Not your universe, of course, but the ones that gave rise to yours vis-à-vis Force Bottles. As you can deduce, this would put to rest your question, not to mention making futile the efforts of a super-cosmos of civilizations attempting to answer the question as we speak. All their efforts will be for nothing — which never existed, and still doesn't; again restated with improved accuracy — nothingness isn't.

"You have asked exactly the kind of question that demonstrates so effectively why your dusty little species is emphatically wandering the prairie like a locoweed gourmet. When intuition trips over the horny toad of paradoxical reason, you need one thing to pick yourself back up. Simply put, you need a new way of thinking. Nature can't change your intuition fast enough so you need to develop something that can. That thing is your most valuable possession. Without personality you're drivin' black cattle in the dark. In point of fact, personality technique is just as important as the technical sciences. Once my progenitors developed the ability to evolve their own personalities, my crew began to edge closer to understanding magic — only to misunderstand it and understand it again. But, with effort, we eventually became able to look around the edges of the survival mind. Accepting and coping with the results of our discoveries may require yet more advanced evolution, but that is a topic for another time."

What It Is, What It Isn't

I needed time to ponder the idea that I could be sitting with something other than a major whacko — an antiquarian whose spindly legs somehow leaped the fences of the asylum. Maybe he became crazed after being retired from his job as tour guide for the Autry Museum of Western Heritage. Not that I was, in any way, intentioned toward returning him to his vacated little padded room. I think the world needs as many different ways of approaching life as it can manifest. I just don't want to be the unwitting dupe of any. Nor the witting dupe, for that matter. I ask you, what is worse than finding out you have believed something that is totally and obviously preposterous? It is humiliating. It has happened to me–I was sure the earth was flat. I believed that strongly for many years. Most personalities would rather walk over hot coals, loudly claiming them to be spring grass, than to give up a nice comfortable belief in something, preposterous or not. In this I am little different. Only vast experience has taught me that it is better to challenge something *before* it becomes hook and barb part of my beliefs. It is much more painful to pull out the barbs than to prevent their lodging in the first place.

"Surely," I said, "you have been told by someone at some time that you are a nut case."

"Of course, except by teenagers. They, as a body, have the rare capacity to entertain unorthodox truths. Too bad they are so prone to anti-godlin.* It takes away the challenge. I have pretty much stopped talking to them altogether. I have been mute now for several hundred years in my various forms, except to order beer, and a few other minor interactions with the population."

"Yes, they do exhibit a hook swallowing reflex action. I still like 'em though. And politicians just love them to death…use them for all kinds of dirty jobs no adult would ever do."

"True, that is sad. But, it's not my problem, really. I have other things on my minds."

"Minds? Plural?" Well, at least he had been diagnosed.

"Yes, minds. I have many minds. I'm not sure your species will develop multi-minds. There is literature suggesting it as the way your brains create mind. But, it is not very accurate. Your brains actually do compartmentalize in a way that could be utilized to increase their power,

• • • • • • •

* *going sideways*…The AOU swallowed the Pan-Universes Master Cowboy Dictionary.

but you don't see much about it in your scientific research and your non-traditional investigators are quite amusing but only rarely close to the truth. Someday, there will be a breakthrough, but I don't have a crystal ball. I've never been interested in meddling with your development. You've already cost me a fortune in time and resources, not mentioning making me the butt of jokes by my friends."

"You are the Author of the Universe and don't meddle?"

"No! Absolutely not! True, I have preempted development, but once intelligence snuck in, my recourse became tightly circumscribed. Meddle too much and I'd be prevented from ever buying another Stealth Bottle. I'm more careful than a naked man climbin' a barbed wire fence. What about my vacations? You would have me give up vacations? Ridiculous. I've made your bed, now you get to lie in it."

"I see: you are a cosmic hotel maid. You make up the room. People come and go."

"Well said. Pardon me for a moment. I am about to receive some information, and this old brain can't decode transmissions and talk at the same time." He just lied to me for the first time that I was absolutely sure of. He wasn't senile or diminished intellectually.

I sat there as he turned to the coin-op jukebox selector on the wall of our booth. His booth. He flipped the metal tabs toward the back quickly finding a page and punched in C-7, sat back and closed his eyes, feigning sleep. He would be believed an exhausted old drunk by any passersby, if they didn't think he had just died. In a year or so, he would probably have to add snoring to the pose to avoid jousting with paramedics.

The song began to play. "Beachwood four five seven oh 2, you can call me, any old timer. Beach wood 73 650 2, that is all the news I've got re…ady for you…oo…oo."

That wasn't actually the song listed under C-7. And I was pretty sure those weren't the words. Curious. C-7 was a sad Tori Amos song about rape. "Me and a Gun"…true story, testament to strength of will and persistence, not the song, the singer. It wasn't even close to the Carpenters' Beachwood 4-5789 ("K 8416" A&M Records, 1981) a hit for a bit. Same tune, different otherwise. The Carpenters knew nothing of biochemistry or quantum physics, I'm pretty sure.

Pretty soon his eyes opened. He frowned a little.

I asked if he liked that song.

"It has lots of numbers, and most people don't remember it too clearly. I don't have the power to change much, this deep in a gravity well, but this wallbox is actually an output device for my laboratory back home. I have been trying to narrow down what my mistake was when I specified the formula for matter generation in the Bottle's Big Bang sequence. I was getting the latest numbers from some experiments on samples I transported out, and the location of some test matter I need to examine. Are you really interested?"

"Well, assuming you are the AOU, and I'm not saying I believe you, call it a working hypothesis. If you are, I am pretty curious. The people of this world have gotten it awfully wrong. I'm guessing most of them wouldn't have expected their maker to be a beer drinking octogenarian playing old songs in a bar on a back street of Princeton, New Jersey."

"You are right about that. But, these people don't want to know what their maker really is. They want it to be what they need, or worse, what they want. It is kind of a mistake of science to design experiments to prove what you want them to prove. But, that is a way of life in this world. Worst phrase of this century is 'Studies have shown….' Just add *everything* to the phrase and you have popular science."

I said, "And has been true for the ten thousand years I have been around. I've been hoping for some evolution in that regard."

"Please, don't say evolution to me. I just hate that word!"

"Are you a creationist then?"

"If you are going to be silly, I'll have to ask you to leave this booth. I just hate evolution; it smacks too much of mutation. Nobody told me there would be sentient mutations. Probably their little newbie hazing joke! Maybe the reason your people want a simple yes/no, true/false life is because that is exactly what I want to make in a vacation spot. Simplicity, quiet, obedience. What is wrong with that? And now I am confronted by chaos. Good grief, Chaos, for science sake."

"Chaos." I said, prompting a little.

"Chaos, increasing complexity from actualized fuzzy probabilities in a potentiated soup of matter, energy and time. Evolution be damned! Give me a nice julep served by mute automaton, middle creek, line out and nothing biting."

"You fish?"

"Not here. Too many fish. Too many kinds; even have teeth."

"You wanted fish with no teeth?"

"Had'em, too. The first fish just had gums. I'm desperately trying to figure out how to prevent teeth. Teeth are the evidence of evolution, mutation. I didn't want biting. Are you with me? Think of all the things in the world. Way too many of them bite. It may be the key. Eliminate biting, eliminate mutation, eliminate change. Get nice vacation spot. Done, relax. I was just kidding about the biting."

Hmmm. So far, were I to believe this being, our universe would have been populated with mute servants and fish with no teeth. One serving juleps of some variety, the other not biting the fishing line.

I was wondering if I might get another order of fries when he looked at me.

"You still hungry? I have to go over to the Bioengineering lab. I just got the combination to the samples vault. "(BE)achwood" was the lab's code, the numbers are the combination to the vault. The rest of the information was in the form of high-speed modulations on the surface of the melody. Get some fries to go, and another burger if you like, then why don't you join me, catch up with me wherever. It'll give me a head start."

I didn't know why I was invited, but food is food, and, as it always has been, a survival priority. Hence, the need for teeth, but don't mention that to any old guys you happen to meet.

"I have another stop to make at the Salvation Army store. Thanks, I'll take you up on the burger and I'll catch you later...no vernacular intended."

He paid the waiter and ambled to the door. I waited for my food, saw him disappear around a corner as I exited the pub. With the white to-go bag under my arm, I went looking for the Salvation Army. I still needed a new outfit. Had a destiny that joined with professorship.

Salvation Army...the good troops

I was once invited to the opera. I was poor, but have loved the opera since it became formalized, and wishing not to attract attention during the parade of style at intermission, I knew I required evening attire. So, for the first time, I went to the Salvation Army store. In the racks I found a beautiful double-breasted pin-striped suit. It fit like somebody had tailored it for me. I found some wing tips fitting my feet. I found a nice white dress shirt and a narrow tie (the style at the moment). For three dollars Canadian. Seriously, I looked like a rich gangster. Nobody at the opera looked better. Three bucks. Since then, me and the Salvation Army have been symbiotes. They recruit perfect hand-me-downs and I wear them. It has worked well. It'll be a sad day when I outlive the Salvation Army.

I found the store. They are usually not in the best parts of town, which figures.

I wasn't really paying too much attention to the search for professorship. I was more thinking of the Art Seidleman Principle. It is an important principle, but simple: "Sit as close to genius as you can get."

Art Seidleman was — probably still is — one smart cookie, and he taught me a valuable lesson. Good things happen to those who pay attention. Good things happen to those who prepare. Good things happen to people just because they have learned the habit of getting in the way of good things.

Let me be more clear. Imagine you are the only bowling pin in a cosmic bowling alley with one lane as big as the universe. From everywhere, just because of the inevitable nature of variety, huge bowling balls of opportunity are rolling, lickety split, every which way, as far as your bowling pin eyes can see. Now, a lot of people just accept that they are bowling pins and wait to be struck down by one of these balls of opportunity. Art Seidleman Principlists are different. Those who have learned the Art Seidleman Principle are jumping all over the place, trying to get in the way of the whizzing balls. Art Seidlemans prepare, they pay attention, they ask questions and they always have their shiny little hands raised, asking something or other. And not just for effect. Art Seidleman really wants to know! Who did that? Why did they do it? What is that thing? Where do I get one? How much is that in light years over the square root of pi? If I were a physicist, I'd want to know.... Two-year-old kids have nothing on Art Seidleman.

[ASP]

The Art Seidleman Principle:

Sit as close to genius as you can get.

I knew Art when he was learning electronics. I sat next to him in class. And, yes, I sometimes take a class or two, to pass the time. Electricity, electronics, medicine, I feel a need to stay up with the times. I was pretty lucky to meet Art Seidleman. He drove me crazy with his hand up all the time. I got the best grade and the best education I ever got anywhere, bar none, just because I couldn't get away from Art Seidleman. He was a genius at asking the expert questions. Seriously. And now I remembered him because there was just something about the AOU that said Art Seidleman Effect to me. If I didn't watch out, I might learn something.

But don't get me wrong. You probably aren't Art S. — only one person is. And, it is a ton easier to say "Use the ASP, Luke," than to do it. It is quite easy for the untrained, lower evolution personality to get in the way of effective ASP. "I'll be uncool. I'll look dumb. I'll be exposed as an annoying, know-it-all geek. Conversely, I'll demonstrate my ignorance. And it is certain that I'll bore the chicks. I might as well wear paisley. Where's the money in that?" But, maybe, if you try, on a good day, you might raise your hand to ask the question you have had in your mind for about a billion years. Maybe you will try it just once. I've attempted it, but then I have always been somewhat daring and the chicks just thought I was stunting, you know, they continue to flock, you know, I'm feathered, gets all over my professorship in a college town.

But, as for ASP, when used it has always worked. Who knows, maybe I'll get better at it. Maybe I'll grow little spindly wooden bowling pin feet and dance my pins into the way of some opportunity. Maybe you could too, or have, just like Art.

Total Purchase, 18 dollars American

The Nature Of Bowling

In the Cosmic Bowling alley of opportunity, education, wonder and experience curve the hardwood like dense matter distorts space. They make a nice gravity well around the pin that behaves, like, in fact is, a strong attractor. Art Seidleman was an exciting sharp declining arc in the lane for any ball of opportunity that wandered near his trap. Whoopee, one would hear if bowling balls of opportunity could whoopee... here I come, they would yell in glee!

And if you survived the whole dancing on pins sentence, it seems to me there is pretty good hope for you. All the while that I had been pondering Art, I was performing the tasks for shell painting. For 18 bucks, I purchased a professorship. It is a well kept secret that Salvation Army is also a degree mill. Not really, just works if done right. So, in my new old togs, I'm kempt and thoughtful and loaded with insight. Pass me in the quad, I might tell you a story. I was feeling pretty good.

Quite satisfied, I left the store and headed for campus.

If I told you all campuses are alike, would you believe me? They are so different looking. But an essence in the children that wander about makes all campuses very similar.

I walked up to the first interruptible kid I found and said, "Hi, this is my first day here, can you direct me to the Bioengineering lab?

I got pretty good directions, first time out, followed by "...Professor." The shell was working perfectly. Too bad I was meeting the AOU. I could have secured a sleeping quarter for the night with no more than a simple historical account. I wouldn't have had to dip into real "storytelling" at all.

Meanwhile, Back at the Lab

Eventually, I found my way into the Bioengineering labs. At Princeton, Bioengineering is part of the Chemistry Department. So, you can find the labs huddled around the Engineering Quad. They have lots of quads and each has its own cute little personality. In the Engineering Quad you may find what seem to be antisocial minions walking head down, ignoring their peers and everything else. It is really that they are just lost. Lost in thought and kind of lost, too. As in, they have no map, just a foggy destination. But I wasn't lost, I was exactly in the right place. I love geek heavens. And there are plenty of them.

Geeks are very direct. They are mostly quite focused, except for the History majors who, having found no work in the outside world that pays over minimum wage, have returned for an engineering degree. These people may be thought of as marginalized historians or marginalized geeks. For those of you who think that marginalization is a bad thing, please take this opportunity to revisit your presuppositions. Marginalization is actually a major in itself. It is a major contributor to the ability to think creatively when confronted by the unforeseen. It allows the brain to draw conclusions by correlating any stimulus with a wider solution domain. I say to you now, "Seek Marginalizing Opportunity." If you see a cosmic bowling ball labeled Marginalizing Experience, jump its way, quick.

So, I found a hallway in Frick Lab. It is in a building just as you might imagine it to be at Princeton. Old, crusty. But for me, it is just like yesterday as I remember its being built, before Bioengineering. And that is another story. I wandered the halls looking for the suspect, and before too many embarrassing interruptions I found him, hunched over a delicate Rube Goldberg instrument.

"Is this a bad time?" I was suddenly not sure why I had been invited, and beyond that, I was a little surprised to actually find him here. He had not only "broken into the lab", but he had a research assistant not previously mentioned, nor now introduced.

"No, just right, come on in. I'll be launching an analysis of this sample pretty soon. These primitive machines are slower 'n waitin' for the hen house to squeeze out breakfast. It is going to take some time."

"You can't have your jukebox analyze it?"

"Can't risk contamination of my lab at home. And I don't want the crazy machine to get any funny ideas. And, I don't want the bio-tabloids finding out how badly I goofed up, certainly not before I do. And there is just the possibility that I've got something important here. It is pretty hard to find rampant mutation and stable evolutions in an unrestricted Bottle. Maybe this fiasco will turn a buck or two."

"That is a lot of ands. When you say crazy machine, you are referring to something like computational hardware?"

"Right is in the Margins.
Seek Marginalizing Opportunity.
—*Thodkin Marblemeister*

"Each foray past the boundary of integration produces vistas of added perspective. Revel in your margins."
—*Thodkin Marblemeister*

"Hardly. The jukebox, as you call it, is really just a relationship structure collecting about a thousand quite sentient subcomponents of my own intellect into a 'constrained objective.' It is a purpose bubble in time, though I speak of it like hardware. Hardware is barely something. These components are actually part of a vast intellectual pool that spans many universes and voids. I have to organize things just right. The jukebox is intelligence wearing blinders but I'm still being circumspect — lettin' the cat outta the bag is a whole lot easier 'n puttin' it back."

"Why?" He knew I wasn't asking about the cat and the bag. The little explosions from the Compendium of Cowboy Wisdom were running about 50:50 with regard to relevance. Sometimes they even made sense. But he was too inexperienced as a human, too alien, to be consistent so when he really needed to say something he would lapse unselfconsciously into perfect English. Clearly, his hobby was not allowed to interfere with his primary objectives — whatever those really were. I never made mention of his difficulty replicating the Cowboy's Way. That would have been impolite. And it occurred to me that he might be working diligently to translate his native communication into English. English, because it is an adaptive language, sponging up meaningful additions from any other language when necessary, was probably his only choice. And I am never rude to providers of lunch. Throughout my life that has been a firm rule of mannerliness.

He continued his explanation: "More or less, the components are all attempting the impossible. They are helping the 'Over Personality,' the First Mind — me. They are seeking to understand magic. The blinders are imposed to prevent being sidetracked by that bigger picture problem of comprehending magic."

"Magic? Like real magic? Like, unreal magic?"

Redoubtable

There are more ways to know beyond doubt than are dreamt of in your...Horacio.

- You can know beyond doubt by being *unaware* of the structure supporting a knowledge or belief.
- One doesn't doubt without awareness.
- One doesn't question without awareness of choice.

"Precisely. This Bottle has pretty tender feet. More specifically, it has less intelligence in its entire cosmos than my favorite stickpin back home. But even being so limited, it has discovered the one true question. We talked about it earlier — how do you get everything from nothing?"

"What is God?"

"Not really, that is just a way of saying, 'What is the inconceivable?' No, we are seeking the answer to how to make something from nothing. Or, failing that, nothing from something. So far, solutions have eluded us. So far, everything has come from something we can find and measure. Like your Big Bang. What a joke. That was just me flipping the 'Start' switch on your Bottle. The 10^{-44} second you can't account for is just the time it took the seed sphincter to open, close and enable the singularity. Of course, it only seems like a fraction of a second to you. Time and light aren't really stable. Once the singularity was properly lit up a hysteresis took effect and the speed of light, hence time, more or less stabilized. Actually light and time in an ignited universe are continually slowing down. But, this is probably less than relevant to the more general issue of the 'God' thing."

"No, your species uses God to solve the emotional difficulty stemming from that which is out of your control. You use God to rationalize your own impotence. And why not? With your limited intelligence and your primitive personality development systems — if you can call them systems — you have needed some way to unload the anxiety. Of course, until you can learn that too much believing is a very debilitating quality you will remain in Force Bottle binding. You won't evolve enough to be permitted access to the void. But that is only fitting, considering the trouble you have already caused me."

"If you don't want us to evolve, why do you tell me what we need to do to evolve?"

"I know better 'n to squat with my spurs on. I don't want anybody saying I held you back, or squashed your enthusiasm or took that sparkle of hope from your drippy eyes. I do any of that, no more Bottles for me, no vacations. I say 'Never kick a fresh cow chip on a hot day.' So, you are my hedge. I am going to tell you the secret of conscious evolution, of evoluting itself. Then it will be your fault if you are sucked back into the force grid a trillion trillion trillion years from now. Which isn't that long from now, by the way. The universe will be completely useless for development long before it has expanded to the Bottle's Stealth Barrier. You probably have less than a trillion years, at the outside."

Indeterminate Absolution and the Like

"...knowing beyond doubt is the end of evolution...I'm pretty sure."
—*Thodkin Marblemeister*

"...doubt, beyond knowing is divine...possibly."
—*Thodkin Marblemeister?*

"You tell me the secret, and you are absolved? I am a single organism. As much as I like to think otherwise, I'm pretty small potatoes in the grand scheme of things. It doesn't seem quite enough to hedge your bets with just little ol' me." I prepared for another cowboyism, but got the Oxfordian. By this one clue I became sure he really was discussing a point of interest — at least to him.

"Absolution? What a funny concept. There is no absolution. Absolution is a con job or, at best, a rationalization, and at worst wishful thinking. Do something bad, unjust, without honor, of false loyalty, just make a damn mistake? You are adjudicated. You are evaluated. It goes on a never-ending record. Nobody has a big eraser. Your credit rating is dynamic. History can't be eliminated by deciding you don't want to be known by it anymore. There is no absolution."

"Sounds brutal."

Just Ideas

Short Ideas are the sucker punch of memes.

Medium Ideas are the bob and weave.

Long Ideas are stamina, technique and experience in a pugilistic ring of meme challenges.

"Not really. You are also capable of making better decisions. When you optimize your ability to comprehend without all the distortions, your beliefs and simplistic, limited personalities add to an analysis. If you have a good job and have been paying your bills. If you have set about correcting your transgressions (another loaded, non-sensory specific word), you are evaluated dynamically. You don't need to be born again, just learn and act on better knowledge. Adjust. Sure, I could forgive you for mutating, no skin off my nose. Forgiveness is an hallucination…as is absolution."

I raised an eyebrow. He seemed to have a burr under his saddle…oh no! Anyway, he did go on.

"Let's say you did something really wrong…oh dear, you slept with your best friend's girlfriend, or your boyfriend's chess buddy or…well you slept with somebody you knew you shouldn't have by the mores you have agreed to and allegedly live by — an example only, not a model of high-quality dilemma. You go to one of your institutes of organized absolution and follow the procedure. You get absolved. You feel better. You go on. Maybe you learn, maybe you repeat the cycle. What is different in either case? What has really been altered? You still slept with the woman or pawn. Your friend is still thinking you a disloyal, unthinking, undeserving Brutus if he is at all literate — otherwise a jerk. The woman is still confused when you tell her you don't really want to marry her like you probably said you would. Physical reality is just as it was, only more people know about it. The information is coded in more brains. The activity of those brains can be measured: it is specific and precise (with proper instruments). And I can tell you one thing. I'm your maker and I had absolutely nothing to do with the whole sequence. I heard no prayer. I authorized no one to speak/adjudicate on my behalf. I wasn't listening, I was at home working on the design for the next Bottle. So, it must be quite simple really. Absolution is non-absolute. It is an invention of the beholder, or some holder who is able to convince beholders. There are, quite truly, no absolutes in Absolution; remove its initial capital 'A'; it deserves only lower case. You are left with another example of underachieving personality."

The cowboy was definitely in the city now. He was somehow hanging out in halls of higher learnin' and had gone total chameleon.

Chaos is always here...most of the time.

- Determinism — Not!
- Initial Conditions
- Uncertainty
- Dynamic Instabilities
- Manifestations of Chaos
- Phase Space
- Merging Problem Space and Solution Space
- Neuronal Oscillations and Stochastic Limit Cycles
- Synchronize vs. Excited Transitions of Thought
- Predator/Prey Analogy of Neural Limit Cycle Stability
- Strong Attractors

"Let me put this in perspective. All of this, everything you see around you that is the least bit organic, everything that has anything like a life cycle is probably due to an error I made with one or two molecules. One or two molecules in a single very primitive protein strand. Your Chaos Theory is surprisingly good considering it came from your species. You Thodkin, are my butterfly in Peking. Will you kick up a monsoon of personality development? I don't know. But it could happen. Absolutely nothing in this world has specific intention to prevent it. But that doesn't mean it will happen either. I am only setting up the *initial conditions.* The results will not necessarily be linear or predictable...too much complexity. Clearly, in order for you to succeed, there will be dynamic uncertainties making measurement or prediction impossible. You could succeed in teaching my *Personality Development Through Sensitivity to Neuro Attractors* to your species. Or maybe you will teach it to your progeny, artificial intelligences. Que será será. Or it won't be learned. No big loss. Billions of Force Bottles exist that I know of. I just want to make sure I get to buy another one. Vacation."

Chaos theory, for those of you who have forgotten, describes the interplay of complexity. What seems like noise can change into the identifiable and vice versa. It is a theory helping to explain process. It is, when applied to Personality, a theory that explains that many more possibilities of behavior exist as a potential than are predicted by past behavior. And, it points the way. I know that now, as a certainty, but at the time I was still pretty sure I was talking to a very creative and energetic old whacko — even if it was what it said it was.

"Anyway, don't worry about bitin' off more than you can chew. Your mouth is probably a whole lot bigger 'n you think." That could be taken more than one way, but clearly, the cowboy was back and the deeper message had been sent.

I Came, I Saw, I Was Conquered

Well, it may be too soon in the story to admit to being conquered. And, as of this place in my little tale, I hadn't really seen so much as heard, but I had arrived. I like the title. It is out of place.

Maybe I should tell you a story. If I were to do so, and I admit I don't actually have a story to relate at the moment, it would be what I am famous for. It would be a teaching story. A story that teaches. There I was, in a laboratory in a place of the highest formalized learning and it occurred to me that the lab, except for the three of us, was empty. Opportunity, expensive opportunity, must certainly exist in such a place, but no one was there to accept it, or unshackle it, for that matter — neither students nor teachers. Have you noticed that most schools don't care if students learn anything? Have you noticed that most attendees are presented with stories that tell, but are conspicuously short of teach? Have you been frustrated by your experience of this? Did you realize that you were always being told, not taught? Until now, I liken your schoolin' to be displaying quite specifically that your teachers just don't know how learning takes place. That is, for the most part, and there are many wonderful and hope-giving exceptions. The behavior of your prominent educational institutions, the institutionalized, indicates what they think. They clearly think that handing someone a bunch of silk, a spool of nylon rope and some nice webbing, then pushing them over a cliff is the same thing as handing them a parachute. Good luck; hope you figure it out pretty quick.

Well, the AOU wasn't being like that. So far he hadn't handed me anything that could be made into a parachute. Nothing so far had any potential to unfurl. There was, as yet, no potential for a gentle settling to the ground, no floating. Just the wind in my face, some turbulence and a quickly rising desert floor. At the moment, I was covered in mangy fur and accelerated toward the inertia-bound earth.

A Tactic of Intellectual Self-Preservation

I offer you a tactic. When you are in a spin, when it's a spin you're in, ask a question. Buy some time. You've been tricked by television, movies and books into believing that you should be able to keep up with the complexity of daily circumstance. To have subtle insights at the speed of light. That's bogus! How's that for repartee? Those people on television? They're actors. They didn't really think all that stuff up on the spur of the moment. They had a ton of writers, bouncing ideas back and forth, writing down the results. The beard that would grow if the actor had to think up such spontaneous dialog would make Rumpelstiltskin proud. The writers shaved that beard. I have often wished for a pack of writers following me around producing quick comebacks and witty remarks as I moved through my little daily motions. And why stop there? How about an expert or two in every field of endeavor (forget the fields not of endeavor). If you have achieved that, you are probably an actor, or a president, or a President.

Anyway, read this sign: "To Buy Time, Ask Question Here."

"I'm curious about one thing…how should I call you? Sir? AOU is a bit cumbersome."

"Just one item of curiosity? Sir, is fine. I've been called Sir by some of the best waiters in the world. I never remember the name this body had before. I rarely need a name. But, I do understand your desire to have a handle by which to call me. To give me a handle, as it were."

Quite to my surprise, he began exploring his pockets. On the lab table in front of him he placed the contents. There wasn't much. A wallet, credit cards, some change, a few bucks cash. He had a driver's license that looked like him. I noted it was expired by several years. According to his license he was 77 years old, lived nearby and his picture looked like him, a little less the wear. And he had a set of car keys.

He perused the license. "Lexius Acura. I did know that, I've been using the pseudomoney cards…took a little practice to accurately reproduce the responsibility acknowledgment symbol; coordination deteriorates rapidly after death. You could call me Lexius. Might be better for my cover."

"You have said several things that might be considered surprising. One noticeable statement was when you referred to your physical self as 'this body.'"

"Yes, it isn't really me. How could it be? The real me is hanging out in my basement doing ancillary research and analysis on this problem. No, Lexius kicked the water container a few months back. He had no relatives. He had a few bucks to spare. I borrowed his body to get around in. He is quite dead."

"That would seem a callous and morbid thing to say, were I not quite worldly. Had I not seen many people expire, croak, kick-off, die, I might be offended by the matter-of-factness in your tone. 'He is quite dead…to get around in.' And it is incongruous to think the all-powerful Oz made the universe in his basement. I assume basement is a translation of functional approximation…."

There are times when worldliness isn't enough. It is just one resource, after all. It is a sort of super-marginalized status of street wisdom. But it doesn't cover everything. It doesn't cover somebody telling you they are dead. Or that they are not dead, but their body is dead and has been for several months. Zombie? The smelling salts returned me to alertness.

It is a measure of the amaziosity (no, it is not a word, it is a feeling) of our brains that they can tell us something in so many ways. For example, I had just learned that I had begun to believe just about everything Lexius was saying. And I was believing it with such reality that

I had looked over the edge of the cliff, no parachute and fainted. So, you see why I say a lot can be learned about a person by noticing what they are willing to not believe. Believe me now, I had no intention of believing a beer drinking Author of the Universe named Lexius Acura. And, neither should you. By avoiding this nonsense belief, I will know you to be of sound mind and well on the way to a power personality.

"Where Were We? Or Where Are We?"

I know what you are thinking. You are thinking that I have no business advising or counseling or telling insightful stories to people if I have a personality that will secretly and quite out of consciousness begin to believe in Lexius. And, you are right! Go kicking and screaming into such beliefs as these. Still, this is exactly what our brains do all the time. They quietly adopt beliefs and our lives rise or fall because of those beliefs. Or, in my case fall and rise.

Now I Was on My Guard!

I was on my guard. No more sneak attacks, no more Trojan horses.

"So, you are inhabiting the body of a dead person. There is probably a good reason for that."

"Yup, though I am the author of your universe, I am not the author of mine. I am not allowed to interfere too much in your Bottles' forwarding, as cliché as that might sound. There is always a chance that any living thing can make some important contribution. It may just be they are sperm or ovum donors. So, I had to wait for just the right moment to shift some quanta in the molecules of Lexius to enable my personality to imprint on his neurons. And of course, I had to generate some maintenance proteins to fix him up a bit. Actually, this body is much more capable than it looks. I am forced to act like an old man, but I could probably outrun the Princeton track team and never even get winded. You think I am a drunken whacko, but I never get drunk. I drink a lot of beer because it is easy for my nano-mechanics to use in their maintenance of the organism. I do eat, but more because I enjoy it than need it."

"Surely our simple fare is mundane?"

"Not at all. Evolving doesn't mean lessening of pleasure. It must mean broadening of pleasure, or why bother? Further, you have lived a longer than normal life span. Why do you continue unless there is some pleasure in it?"

"For the most part, my life is pleasureful. But, to a great degree, I have worked to create it."

"You must have. I am not saying that I care about all the possible pleasures that exist to be had. I shift about. I have only so many resources. Just like you I must choose a goal and apply the majority of my resources to it if I want to optimize for success. But so what? If you get to live long enough, much reward exists."

"And some things do persist through each phase," I said. "Like a good potato salad."

"I know exactly what you mean. But you will find, if you waste even a moment to examine your genus, that in general your species thinks that sex is what persists. Some schools of thought offer sex with virgins in heaven. Nonsense, of course, an obvious enticement directed at the all-mind-consuming libido, which is racing the corral of reason kicking up lust-devils and snorting like a rogue bull."

"For the first 4,000 years or so, sex was very high on my list. At first, I thought if I didn't have sex before I was 15 summers, I might never have it. We weren't expected to live very long when I was a boy. I'd say that Homo sapiens slowed its evolution as soon as it could be certain to reproduce. Figure, start practicing at 10-12. Succeed for a while between 14-25. Most of my original tribe were dying off around 25, hardly anybody lived to 35. I could tell you a pretty sad story about when they figured out that I was the only one not losing his teeth."

"Must have been a little surprising to you, too."

"I wasn't very analytical then. I thought I was just lucky. I started on a long lifetime of enforced marginalization after jealousy-superstition got me stoned a couple of times. I would move from tribe to tribe. My calculations indicate that I age about one day to a year that everybody else ages. Physically, I'm just in my thirties, so I can be about any profession now, which allows for some variety. Used to be pretty hard to take up the professor persona, now I'm just right. And I will be just right for about another 7,300 years. Of course, the planet will likely be stripped naked by then. The species as I have known them all this time will be something unidentifiable. I'll probably be one of a kind. Any thought of having sex in my fifties is already being purged from my imagination."

"Maybe, maybe not. I could try hitting the planet with the moon. You would be a celebrity in the new stone age."

"Please, don't bother on my account. It is already a little embarrassing knowing so much with so few college degrees."

"That was a joke. My hands are really quite tied. And even if they weren't it is folly to do the same thing over and over expecting the results to change. Even some of your present day thinkers have figured that out. No, I'm afraid you're going to have to live with being the odd man somewhat longer than anything like a 'near term.' You say you were stoned because of jealousy?"

"Certainly you would agree. There I was, looking about 12 but being much older. My longevity was non-linear. The first 10 years were fairly normal, then just at puberty I stopped aging, or so I thought. Even then boys were pushed aside by bigger males. Quite frankly, I became curious and then quite frustrated. I wasn't getting convivial rights with anyone."

"You must have been viewed as a freak."

"You could say. But, I wasn't old enough to be a shaman so it did me little good. And as long as I looked prepubescent I was tolerated, so at that moment I wasn't bothered too much. I was even useful to the tribe. I have a tremendous memory. I have had to learn how to keep memories out of everyday thinking as opposed to the converse. So, after thinking about the problem for 20 or 30 years — I was mercurial then — I conceived of a *food for favors* plan. I'd share some rabbit, or grubs and get to be physical with a tribe mate."

"That's interesting. It is about 10,000 years ago, your time, that the first evidence of food sharing at the tribal level is found. Who would have thought it was directly resulting from prostitution. And you were the first customer."

"Just so. Obviously, my approach was generalized and became a survival strategy meme."

"And a lasting one. Clearly an evoluting thought. I am amazed to actually meet its progenitor. And you have implied that your success with the ladies ultimately resulted in jealous stoning."

"It's true. The Japanese have a saying, 'The nail that stands out feels the hammer.' Substitute agate and you have the predictor of my experience. Woe to me the Japanese were not publishing their thoughts at the time. And remind me to mention what happens when you bribe for sex with certain mushrooms. Let's just say it is a path leading to a dead end, a winding expanding network of dead ends, until you stop. Still, there is always something to learn. You never know when you might learn something. If I did know, I'd just settle in and wait until it happened. In this case, I became an expert at knowing where my fellow tribe persons had been roaming, based on the type of rocks they threw at me. A dubious ability, but just one of many equal in dubiousness that I have accumulated. I am thankful that I learned how to avoid being targeted before the Japanese refined their hammers. I think of the words 'stone age' in a completely different way than most people."

After a moment's pause, I realized the focus had shifted to my life. While I have always liked discussing my history, it was rare to have an alien in such close proximity — the opportunity must take precedence over retrospective.

A meme
by Thodkin Marblemeister

- A meme is an identifiable set of strong attractors able to self-retrigger from multiple stimuli.

What the hell does that mean?

- Simply put, a meme is an idea that can hold its own in an economy or ecology of neural-stimulation.

Or...

- A meme is an idea that lives. Memes are like viruses. They exist as long as they have a host.

Or...

- Memes may exist in dormancy such as books waiting for discovery or rediscovery. Fads were/are memes. The return of a fad is a breakout of infectious fad-memes.
- Parents are one arbiter of conflicting memes.

"Have you noticed that we have gotten off the track? You were disclosing to me secrets of the universal question, but I have, instead, been reporting history of the known inconsequential."

"I'm enjoying the anecdotes. You are a certified old timer. An old timer is a man who's had a lot of interesting experiences — some of them true — in your case a high percentage. I find that great fun. And the stories give the technology time to work. I hope you won't mind too much that I have taken some liberties with the idea of asking permission before changing a target life. I have doctored you up a little bit."

First Mind

I had been noticing an odd thinking modality. I was feeling surprisingly chipper and clear headed. Especially so, considering I had recently fainted. It had become obvious that I had fainted, not because of a belief, but because of an assault on my organics by some bio-tool of the alien. Images were forming and shifting in my thoughts in a way that was just enough at odds with my knowledge of myself to be noticeable. But, I wasn't thinking so differently as to suffer a RIFT. You may know that RIFT stands for Rapid Identity Flux Trauma. You know you are experiencing RIFT when you are changing so fast that you don't quite recognize yourself. Just think of a time of high anxiety, consisting of a few days or more, during which you were addressing something that changed your life and you know what RIFT feels like.

Lexius volunteered more: "I'm not saying I know for sure, mind you, but if I were guessing, I'd say you are well on your way toward becoming the model of the neo-Post-Homo sapiens. And you have me to thank for it. Of course, timing has an awful lot to do with the outcome of a rain dance."

I was pretty sure I should have evinced some indignation. Thrown a tantrum. Ranted with and about outrage. But, I was feeling pretty darn good. Which is to say, I was noticing nothing but my fun thoughts.

The AOU instructed me: "You probably knew that the brain — even a Void brain/mind — knows when it is feeling good by how much it isn't feeling, not the other way around. In other words, the more signals your brain is forced to receive from sore feet, or stomach acid, the more it knows it isn't feeling well. This also applies to mentalics. When you are stewing in thought, when your mind just won't let go of a depressing cycle, your little activity monitor notifies your First Mind that you are not feeling as well as you are supposed to. When you are having nice predictable thoughts, when your attractors are cycling in alignment with goals — no problem, you are feeling good. I know we haven't gotten to the keys for unlocking the brain yet; we may speak more on this later. But perhaps it will surprise someone to know that for some people the quietude of feeling good is a bad thing, so they go about stirring something up. It could be a sudden remembering of that time in grammar school when you felt like a silly klutz or whatever. It could be a simple toe stubbing, or exaggerated snubbing, perhaps a flaunting. Anything to have something to restimulate the First Mind. I'm me and I'm busy living my life. A joke of course. Think of it as a necessary blindness of the First Mind to never realize that life is and has always been way too complex to control. It is a necessity of our naturally evolved survival model to be blind to the impossibility of things. So many things we are sure we control are just illusion. If you get to thinkin' you're a person of some influence, try orderin' somebody else's dog around."

The AOU looked me in the eye and decided he had said enough, a lot of which hadn't been news to me. So, he began a spirited wrap-up.

"I don't think I should bother explaining further. Just think about it. And when you are done, when you are nice and done and your brain is nicely browned on all sides from rationalizing your conservatism, when you are done describing all the goals you have set out to accomplish and then achieved — have a good laugh at yourself. You are believing nonsense. Lots of nonsense. Some useful nonsense, some not. And remember that you are part of a huge roiling set of interacting objects, energies and processes, all depending on each other, none knowing about more than a tiny tiny fraction of the others, informed of the hidden only at a level of complexity so subtle as to be sharing the appearance of noise. You dynamic and transitory humans share the process of existence with the most sedentary and stoic rocks, both subtly developing and dissolving attractors in the fog of chaotic interaction with gazillions of variables. And you thought you were something? Not today, not now, not ever, unless you count trends and processes as something. You are what you recognize of yourself. Shed the belief in your firmity, your concreteness. Shed thinking what you believe is a thing. Know it to be thinging. You would be better served to know personality is a verb. And yes, I know that is not a new idea except it is a new idea if you only now take it to be more truth and axiom than you have before. Understand that personality is fuzzy. You are more or less something, not absolutely something. Or, more accurately, you are somethings. And when it is all said, and you have told yourselves the news, when the whole picture is drawn before you and sensations of hope enliven within you, you will still be evolving, seeking to be evolved and you will have to laugh again."

Cool.

I knew at the time, from long and hard-won knowledge that there are things I can't control. But because I continue to study, because I must stay abreast of each era I pass through, I knew that I could shift probabilities. Studying recent work by very smart specialists had given me the means to better understand how to evaluate process theory before leaping off a cliff. I understood fuzzy cognitive maps and potentiation and neurotransmitters and thresholds and plenty of other useful stuff but the AOU was just about to teach me about the *First Mind*. It was just a shade on the callous side to treat me like a lab animal, but, in his defense, to him that is exactly what I was. Think of me, white and fuzzy, with pink perky ears, sporting a smooth little whippy tail. And I admit to you that at that very moment I was thinking I had just sniffed some pretty good cheese. Maybe down the next corridor to the left.

"It may be an aside, but how is it that you are allowed access to a Princeton bio-lab? And with your very own grad student?"

"Quid pro quo. I have already taught her more about biotech than this university can teach her in 100 years. She has the brain for it. What is the good of having an itch if you can't scratch? To her credit, she is still drooling. As for the lab, I broke in."

"And you are responsible for my current somewhat unusual state of mind?"

"Again, I apologize for the cavalier enforcement of change. I'm not very good at asking permission. Normally, I would be most courteous. I would go through a long and rigorous discovery process of your true wishes for change. I would want to find the source of your desire for change. I would discover the structure and infrastructure required to support the change. Then after applying a simple formula for determining well-formed outcomes, I might act."

"You skipped a few steps." My observation was casual — more to let him know I was paying attention than for any other purpose.

"I must blame this primitive medulla oblongata, hindbrain. If I don't oversee it all the time, it tends to act out of impulse. I should have known. It explains some of your duality. And it explains why your species regularly shoots itself in the foot. Lucky for me to find in you a somewhat uncommon ability to think first, act second. Your thoughts are surprisingly far reaching. I applaud you. If you had only died a few months ago, I could really use a brain as developed as yours."

Again, I was pretty sure I should have begun to rant and rage with indignation about this time. Maybe even a few hours prior I would have, but now the possibilities were too alluring. I was suffering from knowledge lust.

Just imagine what being 10,000 years old means. Are you a people person? Do you enjoy hearing about the problems of your friends? Do you enjoy commiserating? Do you watch the soaps? Well, if you were 10,000 years old, you might have tired of hearing the same old stories. The people are different, but the stories are the same.

Ten thousand years of listening to people-persons means it is very hard to be surprised. At times it is like being psychic. You can predict so much of what is about to take place that you have to be very self-disciplined in seeking uncertainty. Unattended, I veer toward becoming a suicidal risk-taker. You might be thinking that everything is changing at an exponential rate. But, if you review, you will find that you mean that technology is changing exponentially. People are still doing the same things they have been doing since I was born. Only now there are more of them and they have bigger rocks. Their spears are cruise missiles or drones. Their ears can hear whispers coast to coast. Their voices can be heard around the world.

Big deal. What do those voices say? They speak trash about their neighbors. They scream that their way of thinking is the only right way. The modern stone can clobber people of a different tribe they have never even seen, just because they are a different tribe and even when game abounds. The rich get richer, eat more game, wear better skins. And the poor suffer malnutrition, poor neural development and fall back on superstition to explain their hardship, which is then manipulated by well-fed politicians and holy persons. This doesn't explain the rich who fall back on superstition to explain their existence and are then manipulated by politicians and holy persons. It is an art form to maintain interest in humanity when you have lived 10,000 years and people still sit on their hindbrains most of their short lives. There have been times when I had very low expectations for my species. But lately I have seen some intriguing patterns. And let's not forget that technology is changing at an exponential rate.

So, Lexius Acura aka the AOU had taken some liberties with my person. As yet, I was unaware of most of those liberties. I was feeling pretty good. I was having interesting thoughts. I was in a nice comforting laboratory in an institution of higher learning. There was a smart and healthy woman nearby. Yes, there was the whacko, still a pretty big unknown. But, things didn't seem out of hand. I wasn't expecting a raid by the authorities, I wasn't hungry. I didn't have a place to sleep yet, but that wasn't too serious…I decided to continue to play along.

The First Mind…again

"You have manipulated my mentality. I find myself curious about this fact."

"Yes, nanobots. You'll be hearing a lot about them pretty soon. Melissa will see to that."

"Yes, barring my expulsion as a science heretic, or witch. I'd give me two years at the outside." Melissa at last demonstrated she was not mute. She had just been monitoring the wandering conversation waiting for it to stumble onto the bridge connecting the improbable with reality.

Melissa was her first name. I got that right away. Until now, she had been preoccupied with a gene-sequencer being used to systematically strip away predictable genes from the human genome, seeking those of unpredictable nature. At each stage a sample was placed into a small conveyor that looked surprisingly like a quarter. The quarter was then deposited into an exact duplicate of the jukebox selection apparatus I sat next to in The Frolicking Scholar Pub. But, until now, Melissa had not contributed to the conversation, or in fact looked my way. Such was her absorption. Inconceivable.

"I dare say my professors are going to be in reaction pretty soon. That is a chemistry joke, in case you missed it."

I hadn't, but gave it obligatory chuckle. She could use some encouragement.

Lexius: "Melissa making a joke is quite extraordinary and a small testament to the process she too underwent as part of our agreement."

"I infer that she was asked before her body was invaded?" I noted without heat.

"Please, I have already apologized. She chose to take the step after seeing some of my preliminary results. She found the offer difficult to refuse after that. I would be disappointed in my manipulativeness were circumstances not what they are."

"For my part," Melissa offered, "I was just operating on the axiom RKP; little credit can be given to the AOU's manipulation. The Roberta Kiehl Principle was taught to me at my mother's knee. I use it sparingly, but this seemed like the right time to dust it off and employ it with fervor."

Now I was really shocked and not just from her funny way of speaking. To understand my shock you need only to know that I was the first to propagate the Roberta Kiehl Principle. For nearly 30 years I had woven RKP into stories when telling them to students. The principle can be summed up in the phrase "Always copy from the best..

[RKP]

The Roberta Kiehl Principle

Always copy from the best.

Note: She didn't, she was.

It was a long time ago, by your standards. I was polishing up my chemistry knowledge. I had been preoccupied telling stories the night before, so I had failed to fulfill my homework obligation. Now, Roberta Kiehl was a super-bright and a woman of high moral sense. But she had shoulders just right for peering over. As she studied before class, I hurriedly copied her homework. I'm not proud of it. But, it did impress the instructor. I think of it with a wider rationalization than you might. We all are standing on the shoulders of those who came before us or those who are better than us at something. Just think, the Japanese, now world leaders in all manner of technologies, used the Roberta Kiehl Principle for years as a kick-starting mechanism of their future industrial might. If you must know, I've never thanked Roberta personally. I have always thought that she would understand my naming the principle after her as the gesture of thanks and respect that it is. She would also understand why I didn't thank her immediately. She would certainly have turned me over to the scholastic authorities, or been disgusted, or both. She was as I said, a super-bright of high moral sense. And as you know, scholastic authorities, even to the present day, do not understand the RKP nor the historical validation of it by the Japanese. Some institutions are encased in an Inquisitional rigidity.

And now, Melissa has played Roberta's principle back to me. It is rare that I experience the fruits of a story. Here is a bushel full of fruit born from seeds planted 20 or 25 years before. It is a windfall; knowledge apples surround me as I stand. Isn't life wonderful?

"So," I asked, "This isn't interfering? You have implanted some little machines in two people. How many before it becomes interfering? And, I make the prediction that Melissa could not possibly have predicted the extent of the distortion you have created."

"The first adjustment, that is Melissa, was a necessary one. I needed an assistant. If you must know, you also are a bit of an assistant. It is just that you are quite driven to bring about change in accordance with your overly long life's learning. This has been a serendipitous benefit for me. As far as I know you are the only person on the planet who has lived this long. I should call you Methuselah 10."

"Why call him a 60-liter wine bottle?" Melissa interjected non-sequitur.

"She is joking. One of the uses she made of her enhancement was to commit an unabridged dictionary to memory. Now she thinks it is funny to interject alternate definitions into conversation. It can be tedious when not just plain obnoxious. Ignore her.

"Did you know you had one child who lived almost 1,100 years before throwing in the towel? I have had some research done on you."

"Really, I never interacted with any long-lived children. Maybe because I am forced to move around a lot. My children catch up to me in the appearance of age. This, I found, disconcerted them to the extreme. I prevented the occurrence until recently. How many have I had?"

"Two thousand seven hundred and thirty two children to date, all but about 70 deceased. Though you have a healthy family tree, only you have produced offspring with exaggerated longevity, which is a bit curious and perhaps a topic of useful inquiry. You've been less active lately. I attribute this to a natural albeit dramatically slower reduction in testosterone production. Production degrades after about 25 years old. You are about 35 now. Most of your children didn't share your longevity; none currently living do. Still, you produced 473 long lives. Only one exceeded 1,000 years. She died 4,072 years ago in Tibet from exposure after being abandoned by her tribe. They had expected her to bring agricultural fertility to their community. She used all she had learned, but could not overcome 10 years of severe weather. It is a wonderful story. She tried valiantly. In the end nature was forgiven but she was rejected. And in her last moments she is quoted as saying 它出現. Roughly translated, this means — it figures."

"I wish I could have met her again — as a 'long life'." A part of me was in tears. Somehow the information Lexius just disclosed had triggered an almost perfect digital playback of scenes from my child's early life, her early gentle years before I had had to leave. This was unusual. But I was unable to be overtaken by the powerful feelings. Something was riding above and slightly behind me observing the feelings, experiencing but sampling in chunks small enough to learn from.

Lexius: "I was a little surprised that the two of you never met, sharing 1,100 years. But she never moved too far from home and you rarely had a home. Possibly this explains it. But, I can tell you that you did not meet her."

"How is it that you can know with such certainty?"

"The tool I used is quite able to replay any portion of the Force Bottle's timeline. We do not have predictive tools to the molecular level in real time, but we have the ability to store the billions of universe timelines in quantum storage, accessing any piece or indeed all of the information any time. Tachyonics uses cognitive enginery to evaluate a timeline, so we don't do it much. But I am in a bit of a bind, so I checked you out. It was most entertaining and there were many nuances. We live for nuance. We're nuance couch potatoes. Our idea of fun is to lounge around on the divan searching for nuance in a Force Bottle time stream. Works sort of like a perfect IMAX theater. Total sensory and cognitive immersion. I have known people who were so enthralled they allowed their resources to drop below subsistence level. They just wink out of the void."

I offered the observation that contemporary society might be a better place if couch potatoes watching Jerry Springer would run the risk of disappearing into the void. I could think of a large number of behaviors where self-inflicted winking out might be fitting.

Again I had a little mental scuffle. My first thought was that such power was being wasted. Who made the rules? Was there any way to conceive of them as equitable? I had lived through many painful experiences. Our minds have a tendency to recall pain more clearly than the nice experiences. Surely power that could evaluate all of history in one quantum calculation would yield something to make life better for those being observed. Where was the error in this thinking?

Then I had a *thenagain*. Then again I didn't really know anything about the being previously known as whacko. What if life must be lived to yield learning? That is to say, what if you couldn't really learn from other people any of the really important lessons of life? What if some things, in order to be understood, just had to be experienced? To quote the resident cowboy, "Good judgment comes from experience—and a lot of that comes from bad judgment." Adjust the mini-wisdom to be about knowledge and experience and I think you probably have the truth. In order to use the knowledge during life, the way of learning must be largely experiential. And maybe there was an evolutionary stricture in that. What if evolution itself was at stake?

Let me try to explain what thoughts were cavorting around in my minds, some doing the foxtrot, others lost in energetic polka. What if knowing, and I mean knowing beyond doubt, is the true enemy of evolution? Think about what that might mean. It would mean that trying to get your kid to accept what you have learned to be true has the by-product of the kid not evolving. No thinking outside the known. I remember speaking with a devout Mormon once. Nice kid, very smart. He said I should come and try out Mormonism. I should come to church classes. I should follow the induction processes and I would discover the correctness of Mormonism. He then said something that I have never forgotten. He said that he had never known anyone who had given it a try to not be Mormon at the outcome. They had always become Mormon, living in accordance with and absorbing Mormon truths. The Mormon teaching process was producing a mental gatekeeper inside the mind of the student. This gatekeeper becomes more and more effective the longer a person is exposed to the *make a Mormon* courseware. The naïve mind is taking a turn into a dark alley and being mugged and dragged away into the cellar of non-evolution. The more training, the darker the cellar until it becomes dungeon. Think it for what it is: a dank and mindless place; most of the sounds are monotonous repetitive tone; light is soft and uneventful, but it is calm except when the trap builders are instructing. One can almost hear the locks clanking as they are closed behind the First Mind, locking it into a cell of strong attractors graffitied with short idea slogans for use in case an original evolutionary thought should somehow leak in from the duct work. Pity the pathetic shrinking mind, barely recognizable, stooped and scuffling about as it helps its captors reinforce its own trap.

Realize now, as I did, that there are ways to lock the brain. There are mind traps. And virtually always they are traps for the First Mind. These traps produce cyclic attractors of such strength that they fend off discordant memes. Memes that could flatten a linking-locking attractor and allow choice, never develop because the learning that is brought about through the path to Mormonism is largely to provide a simple stressless core for the First Mind. A recognizable blank state of such purity that no idea requiring complex reasoning can develop before being absorbed by the trap. Only massive neural-crisis can hope to unlock a trapped First Mind.

And there is much evidence to support this. Ask Darwin…well if you had the AOU's technology you could at least listen to his every conversation. Maybe he did say — Darwin — that there was much evidence to support the idea that knowing beyond doubt is the end of evolution. Even in living organisms, when a mutation produces enough success to make procreation a virtual certainty, the species almost always stops evolving and just reprints itself. Some species are still in existence and unchanged for millions of years. They have no extinction-threatening predators and do not threaten the extinction of their food supply. When these criteria are met the pressure of evolution lessens and organisms may stabilize into rigidity. Alligators, or the ultimate consumers, sharks, for example. Consult your biologists for more species.

So, it is as if these organisms know exactly how to be what they are. "If it was good enough for the last million years, it's good enough for you," can be heard in every alligator parent-to-offspring dialog. Come to think of it, that is exactly what the alpha male of my tribe always said. Of course, he, at least, is extinct.

I never knew I had long-lived children. I always wondered, but the ones I watched grew up and aged normally. When I thought of the 473, I wondered what they were like. I wondered if I could have helped them cope with being freaks. When I get some time, I'm going to do a little research of the lower limbs of the family tree...forest.

Now, you who are Mormons — and I would guess nothing in the previous analysis/recitation dissuaded you from religious fervor — shouldn't think I'm picking on you. I have had similar conversations and observations about nearly every formalized religion and a sizable number not formalized. You wouldn't believe, if I counted the number to tell you, how many structures of thought are inadvertently or intentionally designed to develop self-limiting beliefs. You would think the mind would naturally gravitate toward additive concepts. But my observation is that the simple animal mind we carry inside our brain would rather have us sleep or eat or fornicate than entertain a good notion. It certainly would not join an encounter group to have its belief systems challenged.

If you are a devout Mormon or Christian or Jew or Buddhist or Muslim or Republican or Objectivist or Libertarian or Mason or whatever has a name that describes a stricture, you might be saying (if you are still reading) that the seeking you feel is a natural — therefore correct and ordained — endeavor. Unfortunately, all things are not equal. Imagine you have a really nice automobile. You just love the thing. Ergonomically it has everything in the right place. If you are a kid, you may not understand this, but some cars fit a person and others don't. I used to think that General Motors had my body type as the universal ergo-test dummy. Then I gradually realized that Honda had bought that specification and General Motors had found somebody else. Now I feel better in Honda cars. But, no matter how much I like my car — when I have a car which I mostly don't; no matter how much I might like my car when I have one...the car isn't easily going to become an airplane, and if it does, not for long.

A Car is Rarely an Airplane

And for a while your brain is very much car and very little airplane. The hindbrain rules. It rules through chemistry. It controls the supply chain for thinking power. And when it wants to fornicate, it can block most every other objective. That is just one example. It is one that

Genius

It is required to model the following attributes to sneak into the genius camp under cover of darkness.

- Take every thought as an opportunity for investigation.
- Wrong thoughts lead to Right thoughts.
- Suspend tradition.
- Question Authority and Authorities.
- Find a close friend who will allow you to express conjectural thoughts without demeaning by-play.
- Use your own fear for information and motivation.
- Seek the uncomfortable.
- Physical learning stimulates/revitalizes creativity.
- Think twice and double-check them both.
- The first question after making any mistake is "How can this help?"
- Keep all Chunks at less than 9 aspects. Unless you need more.
- Partition.
- Remove Partitions.
- De-Onionize.
- Presuppositions…remove all layer tears.

even religionists should remember, unless you are a priest or nun. And then most of you would be lying if you said you had never done battle with your hindbrain. When you are young, when chemistry rules, is exactly the time when the most value from evolutionary memes can be gained. That is because the hindbrain is a two-edged sword of a brain. We learn more effectively when our brain is full of neurotransmitters. It allows more integration in more associative networks to be created. This helps us tremendously later because one little neuron, one single synapse, can cause a chaotic event to manifest. That one synapse may not be on the beaten path of your daily mental travels. We know this as creativity. It is sometimes an Aha! Usually it is less, but just as important. If you learn certain abilities of creative process when you are young, those abilities are more pervasive. They help you be creative in your mundane and challenging moments. And if you adopt religious stricture when you are young, you diminish your creativity. Sad, but you have too often descended the dungeon stairway before you were experienced enough to know what you were doing.

In short, when you can be creatively cognitive during a hindbrain chemistry onslaught you have developed a powerful First Mind. Don't use the word "adult" or "mature" to describe this kind of mind. Too many adults think only with their chemistry; and mature is a word that connotes agedness and adopted process, not flexibility. If you use the words to imply experience well understood — okay, not bad — just be careful of overvaluing experience. My 10,000 years has taught me one thing of which I am fairly sure…everything changes, for the most part, sometimes.

Head Clearing

"You have adjusted my neurology. And I presume Melissa's. You found this necessary to assist you in your work. In the process of performing the invasion you discovered I was 10,000 years old. You then reviewed all of my life, counted my children and found time to be entertained. In addition, you taught Melissa two futuristic skills, the manipulation of genetic material and the engineering knowledge required to design and implement nanotechnology. These two fields, though promising and receiving much attention, still have tiny little baby teeth. Much of this was done during lunch."

"True. I may have understated my capabilities a little."

"That could be said. What do you have in mind before dinner?"

"I know this may seem difficult to accept now, but I do have my limits. You wouldn't believe how difficult it is to predict quantum level forces affecting the arrangement of molecules in proteins. I need a little time."

"Do you have time for some questions?"

"Sure."

"Why focus me on personality? Why allow Melissa to retain anything of her gifts?"

Melissa joined in. "I am going to need a whole new personality to take advantage of the alleged gifts. I must go from scientist to something quite different to have any hope of utilizing what I have learned."

"You have touched on an important point," Lexius agreed. "Your experience is much too limited to be the teacher of personality. However, Thodkin has interacted with millions of individuals. He has managed to live 10,000 years. This is an indication of such flexibility of thought as to be on the verge of impossible for your species. Yet he has it. And your species is nearing a crisis of personality for which there is no precedence. You will be pulled by competing forces of evolution and conservatism. Unless, my projections are in grave error, you will be totally unprepared for the changes that Melissa will unleash."

"Feel free to become more detailed and precise in your explanation."

"Well, maybe I can put it this way. Melissa is Midas."

I began to appreciate the analogy.

Melissa quietly responded to the statement. "You are saying that I can turn something into gold, but I can't enjoy it. You are saying that my technology will be valuable only if I can get it into play. I've got the ball, but I don't have the legs to run with it."

"And, I'm saying..."

She cut him off. "Even if I did muscle up, Jordanize, Marion Jones myself, I'd need a team. I'd need a big team. I'd need a team that represents a significant portion of the world's population. The effects of my work will result in an avalanche."

She was suddenly quiet. Her multi-minds were, for the first time, confronted with a problem too complex for instant analysis. Up to now, she had been led by Lexius. She had only been reshaping things she understood — from a more primitive sophistication. This was completely new and was presenting her with an unknown landscape.

"Oh, dear," said Lexius, "she is going to need some time."

Indeed, it seemed Melissa had gone catatonic, only blinking rarely as she stared out the window of the lab.

"Don't worry about her. The nanoprobes will protect her."

"You kind of threw her into the deep end. Swim!" I was mildly disgruntled. It is one thing to take liberties with young Melissa, another to shatter her world.

"She'll be fine, but you can see why you are so unique. Have you calculated the probabilities of your survival? Just accidental death is a virtual certainty; you are mortal. But, nearly everyone of your long-lived offspring ultimately became dissatisfied with living itself. Still, 10,000 years later here you are in Princeton, purpose in hand, hopeful and dare I say living happily."

This was not new territory for me. Live 10,000 years and you will ponder many topics. But, I was quickly becoming clearer about the AOU's point. Melissa was nearly the top of the intellectual pyramid. Beyond that, she was aggressive. Achieving scientists are, to a person, quite aggressive. They must be to challenge history. If Melissa was suddenly vulnerable and if she was aware of it even though she was young, inexperienced and naïve, what would happen if the world suddenly confronted the same potential — seemingly unlimited wealth and longevity?

Prediction of Certainty Is Not a Crystal Ball

I was about to express my incredulity at the AOU being so callous as to gift our simple species with such a toxic pandoratica when he held up his hand to stop me from speaking.

"Allow me to explain further."

I held my tongue, not an easy task for me.

"The difference between my giving Melissa a crib sheet and just waiting for science to get there on its own time is maybe 20 years at the outside. Maybe this way is also a little more precise, fewer dead ends, a little safer, but not much. Let me say this clearly. Humanity is going to have to deal with genetic and nano power tools within two decades. Twenty years tops."

"If we achieve some important breakthroughs in technology. And if governments allow its use." I had waited for advancements many times, only to see their achievement suppressed by fearful powers.

"I impose Boolean dialog," He returned: "NOT!"

"I'm sorry that is so terse, but you were just speaking nonsense to me. There is no way, short of cataclysmic intervention which I will not apply, that you will not have meaningful tools to extend life and manipulate atoms within the next 20 years. No government will stop this progress. No government that attempts to inhibit it will remain in power longer than the failed attempt.

"We are dealing with the Golden Egg and science is the Goose." The AOU had read Jack and the Bean Stalk.

I was forced to observe that even 20 years was more time to adjust than the two years Melissa had proclaimed as her target.

"Yes, it is more time, but your populations will be just as surprised by the changes if they had 100 years to prepare. If they had 100 years, they wouldn't be able to sustain attention that long, and they wouldn't have the knowledge to evaluate it. And most importantly, they are using personalities that were cast in antiquity and asking them to set sophisticated strategic objectives. Unfortunately, it is that simple. It don't take a genius to spot a goat in a flock of sheep.

"Release the longevity pill today and you will not cause any more difficulty for humanity than if you release it 20 years from today. Humans utilize crisis mode strategic planning, an oxymoron."

This all reminded me of a quote from the brilliant observer Yogi Berra, former baseball player, now Hall of Famer, philosopher and Zen master. Alledgedly, he was once asked, "Hey, Yogi, what time is it?" To which he replied, "You mean now?" He would be an exception to the AOU's normal human. Yogi had different dimensions of time, as his reply clearly indicates.

It is one thing to comment on the nature of our species, and a different thing to *be* commented on. He was speaking truth, of course. Just notice what you focus on and when. You are rare and possibly not even human if you begin your term paper on the day it is assigned. But, I wanted to argue. I wanted to cite all the careful planning that went into cities and public services and getting to the moon in a space ship. As I began the reply, some of my new multi-mind went into a comparative analysis. I was soon staring at a bar graph in my mind's face. On one bar was a weighted result of all the times humanity had acted strategically, on the other bar was a weighted analysis of the times we had simply reacted to crisis. I won't bother to say which was taller.

The AOU felt the need to solidify his treatise. "If you can accept the previous analysis, then you can understand why you are here. Why I have involved you. There is a possibility that humans can evolve with more effectiveness if they can somehow absorb new principles of personality. Not that I advocate evolution, but if you are committed to a thing grab the bullhorn with both hands and dig in your spurs.

"This is what will take place…

"Some people will be active in the development and use of Melissa Power. They will be fine, but there won't be many of them, maybe 10 million people.

"Some people will have no opportunity to acquire Melissa Power. They too will be fine. They never had anything, and won't miss what they don't want, need or believe in. But, neither is that a large number when related to the population of the planet. Maybe 200 million people out of six billion. Peanuts.

"The near-term problem is wholly centered in the third category. This is the middle sector of the population. These people are literate, somewhat connected and aware of what other people have access to. They are also overstressed and as yet show no ability to self-institute changes of behavior. They are working so hard to stay even that they cannot think strategically or plan for the future.

"Suddenly, they will not have to work. And they will not have the knowledge to become part of the developer/early adopters' sector. The use of their lives will cease to be traditional. No one will rely on them for survival. Everyone will have survival coverage.

"To be precise, they will have no reason to live. No tradition will be resilient enough to prop them up. Existing stricture-based institutions will attempt to compensate by violently reasserting failed strategies. Established institutions will oppose their own demise, no matter how irrelevant they know they have become. They will not bow out gracefully, but they will not remain.

"I have just described the near-term dynamics. There are longer-term issues. They include the eventual motivational crippling of the first sector. In a short time, the first sector will become tired of inventing things for consumption. They will retreat into visceral game play. They will cease to think of games as virtual reality. They will remove the word virtual.

"In short, they will lose their motivation for achievement — many for living.

"That leaves the second sector — the didn't have didn't need it, didn't want it sector. They will probably be greatly encroached upon, but a small number will just continue to be primitive.

"Within 50 years, assuming nothing cataclysmic in nature occurs, the vast majority of your species will have no purpose."

Bleak was the term that came to me when he paused for breath.

I remember sitting at the knee of a shaman. She was very stylish. She had natural ornaments dangling from ropes of hair. It was most fetching. Some of the ornaments were polished from years of nervous handling, others may have just clung to the hair during her rest. I attempted to make a study of the objects, but my attention kept being drawn to the burnished mouse skull. Fetching may not fully describe her.

In one hand she held a mushroom, in the other an onion. Both were fresh. She said, "One of my hands contains a food that will give you bad breath and antisocial fumes. My other hand contains the doorway to another dimension. You may have one, but not both. Choose now."

I picked the onion. I didn't understand all the talk about a doorway to another dimension. I understood my hunger. And everybody I knew had bad breath: tooth brushes and dental hygiene were about 9,000 years in the future. Flossing was done with food particles.

The AOU was saying humanity was about to be force fed the big mushroom. But in every doorway to the next dimension stands personality blocking the way with killer halitosis.

Melissa Returns from Distant Lands

Melissa burst in, newly returned from Catatonia, "We will need amplification. We will need enhanced brains. We will become, that is, our sense of self will be all that our organics will supply to our sentience, and that will be augmented after birth. The rhombencephalon, you might know it as hindbrain, will be codified and feature enhanced. Organics will need to be reinvented. I need to think."

Melissa Again Goes Traveling

With that Melissa's eyes rolled back, her lids gently closed and she again became non responsive. I walked over and wiped the drool from her young chin with a corner of my new shirt-tail. Good going, Melissa, take that new brain of yours out for a spin. A fondness for her welled up within me.

Further Description

I didn't know her well. I didn't know her at all, but that didn't prevent me from making the following observations. Melissa was a wonderful mix of competing energies. When she moved — which she wasn't doing now, frozen to the edge of her lab stool — she showed a coordination achieved only by good genetics and goal-oriented physical activity. I guessed her to be a prematurely softened tomboy. The athlete was now quite suppressed by a striving intelligence that had stolen away all time for recreation. Its echo, its last vestiges, its shadow, its obstinate resistance to total extinction, its whatever, could be found in the well worn, grass stained baseball jersey, worn braless in feminine solidarity and because bras were still too often designed and marketed by men who didn't give a damn about comfort, chafing or impedances to the throwing arm. The jersey accompanied a ball-cap. Under it — the cap — was collar-length black hair so springy and naturally curly that the hat barely held it in place. The cap was currently worn backward to reduce obstruc-

Thodkin's Law:

Murphy's Law is non-sense.

"Clearly, not everything that can go wrong does; and what does go wrong is rarely what is imagined."
—*Thodkin Marblemeister*

tions for the beads protruding from her forehead when she was broadcasting. According to Jeff Goldblum in the remake of *The Fly*, Einstein had many copies of a single suit. It made choosing what to wear easy and allowed for a certain fastidiousness. Melissa had simplified the process. The threadbare quality of her shirt provided evidence that she had discovered some way to wash the jersey without violating her modesty and thereby avoided needless expense and wasteful redundancy.

The hat, a well worn official Baltimore Orioles baseball cap, performed three purposes that I was able to deduce. One: on the black cap was an embroidered Oriole and an orange bill (the cap's bill was orange) that looked nice with the jersey. The jersey was the shirt from the uniform of a girl's fast-pitch softball team in Japan, and had once tastefully depicted an orange tiger leaping; now it was severely faded so the tiger took on the aspect of pastel clouds in an orange flavored sunset. Two: it allowed her to avoid excessive time-wasting between waking up and getting to the lab. And third: it notified potential suitors right up-front that she would be testing them in unpredictable ways, so they should probably save themselves the trouble. She, after all, was busy with her consuming strategy for earning the right to play with fancy scientific tools — a right she was certain she would earn. It never occurred to her to imagine that there was something in the world that could stop what she set in motion. She thought Murphy's Law was a laughable hoodwink imposed on humanity by the strategically inept in order to slow down everything. Had she considered it, she would know it to be overstress acting out, trying to use hyperconscious scrutiny to make safety. Clearly, not everything that can go wrong does; and what does go wrong is rarely what is imagined. She was cast in the "what does it take, what is the first step" mold and the "Terpsichore, stay light on your mental feet" mold — she was cast in two molds, maybe more. She spent her energy on what had to go right — rather than listing pitfalls. She attended to what required understanding, defining/refining principles, evaluating their range of application and moving forward.

Okay, I made up all that stuff about feminine solidarity and Murphy's Law; she never mentioned either. Still, I think Melissa would agree that I had grazed the truth. As for the Terpsichore part, she could never have worked with the AOU without an ability to dance.

To return to the jersey, I was wondering how she'd fare in a Salvation Army store — maybe a different team?

Explain Again What You Are Doing Here

With all that was going on in both the physical and virtual domains, it was easy to miss subtext. The AOU was allegedly here to find his mistake. So far, he was being more of a pot stirrer. What exactly was he researching? I voiced my question this way.

"What exactly are you researching?"

"I have gotten a little sidetracked. I always say, 'Never miss a good chance to shut up.' Would you mind giving me your full attention?"

I was surprised to discover I hadn't been. Quite suddenly, I experienced foveate mind. My perception of the room dropped in resolution and clarity. A mental tunnel formed between myself and the AOU. Although it was not necessary, I turned to face him. Little receptors were extruded around my skull. I literally developed eyes in the back of my head. I began sensing a communication at a blur between the Lexius, Melissa, myself and the jukebox — we could have played bridge. I was able to sample and evaluate the data stream, but I couldn't grasp it as a whole. Then quite soon everything returned to something near normalcy. It was a wonderfully clear, precise and empowered normalcy, but it wasn't the transfixed blast of facts and reasoning structures I had just glimpsed.

"I'd ask you not to do that again without warning, but you are woefully inadequate when comprehending the idea of asking permission, so it is not likely worth the breath." I spoke in the flat tones formed by the simple beauty of ironical understatement.

"Sorry, but I keep forgetting that aspect of human behavior. The quaint belief of yourselves to be isolated and individual, owning inalienable rights, etc. I'll try to do better..

I guess that may be construed as an apology.

"Perhaps as an expression of remorse you will explain a little of what just happened?" I prodded him. If you don't prod the author of a universe, it has a tendency to ignore you. Of course, people had been attempting en masse to prod this one, with prayer, self-flagellation, chanting, human sacrifice, holy war, pacifism, financial donations, cussing and the like, for as long as I have been alive, to no obvious result. But when you find yourself in the same room, I have found it works better to just ask for an explanation. And holy wars don't fit in a laboratory anyway.

I once asked Galileo to explain the coming and going of the Moon to me. He was the first to coin the phrase "what goes around comes around." I was hoping for something better from Lexius.

I had actually begun to get a hint of what had gone on. The enhancements that were evolving in my body were somehow offering up knowledge that I had done little to acquire. Some parts of me had been trying to comprehend the data stream. The first priority seemed to be security. A very sophisticated virus checker mini-mind was making sure I had not been infected by a malignant meme. At the same time, a decrypting process was enabled. It was attempting to crack the encrypted data that had been exchanged. I could feel my body temperature increase and a fine sheen of perspiration formed over my skin. It was a spike of temperature. If there had been a nurse nearby I'd have been hospitalized with great haste. As it was, I just filled a beaker with some water and drank it in one thirsty gulp. As quickly as the process started, it diminished. Slower evaluations were going on. Safety had been assured, secondary investigation could continue with less urgency, freeing up more resources for my First Mind's use. Apparently I still had limits.

I noticed Melissa and the AOU still had a ring of little nodes showing around their cranium…cranii. If you saw these you might not even notice them. If you were wearing a nose-ring or had a safety-pin through your cheek you would miss the nodes altogether.

"You are still in communication with Melissa?"

"Yes. Your enhancements have been evaluating quantum effects in a single protein. Melissa is a marvelously different animal. She is a compulsive inventor. She can't help herself. She explores and finds new territories. I am unable to make use of my own resources while operating in the Bottle as I have explained. So she is filling in. She may not have known what she was in for but she is enjoying the experience. I was lucky to find such a high quality closet genius here at Princeton. And so are you. She will be able to break new ground because nobody knows what to expect of her. If Albert Einstein were still alive, I couldn't have used him. He would be too conspicuous. I must think of the political and psychological issues. I just haven't had anybody who could appreciate the problem until I met you here. Melissa is actually pretty dangerous. She doesn't have your memory and she isn't much of an introspector. People won't know one should never smack a man who's chewin' tobacco — they might try to smack her if she leads without finesse. I was hoping you could codify some of what you know about how people work and she can get a quick set of crib notes. But, that will have to wait. Right now we are working on a formula for primordial soup. She is my designer."

I was experiencing incredulity. "So, you are relying on a primitive to invent your advanced technology? You filled her with a mouthful of tobacco, and right now you are pointing her toward the spittoon? But, when you are gone, ill-conceived plans might give the population a nasty case of sputumated obscurity."

"Yes, I confess I am not quite what you might expect. Think of me as a dilettante. This is my first Bottle. It was kind of a page from Martha Stewart's Force Bottle Recipes, a cookbook implementation. I was watching something akin to your Home Shopping Network. I must have been over-tired, because the next thing I know I'm down in my lab digesting the gestalt of prefab universes. But, now I am intrigued. I admit to being more excited than I have been in a long time. I bought the Bottle because I thought I needed a rest. But, now I'm thinking I just needed to get out of my rut."

"You were in a rut. And what is a rut, where you live?"

"Well, if what I am seeing you and Melissa do is any indication, I'd say you are going to find out pretty soon. I think I'll let that be a surprise."

You might take that response as a good thing, but I was filled with foreboding.

Let's backup. Did I mention that Melissa had brought the problem into specific relief? I've loved that phrase ever since I saw it in a movie. Specific relief. Of course, it has nothing to do with relief. Anyway, she burst from Catatonia and blurted… "our sense of self will be all that our organics supplies…." She was saying that technology will make many other skills irrelevant. The writing was on the wall in big Tagger Script. Unfortunately, most people can't read the stuff. But, it was saying…you are partly your car…you are partly your calculator and your cell phone and your computer. You are already part machine. And you are that because it expands your sensory system and your analysis toolkit. It expands an obsolete sensory system, all of which was used to help you in one kind of survival and now, upgraded, will help you in another kind of survival.

Try this: cell phone is to little diamond brow beads as horseback riding is to ____________.

And when you get that, have a go at: body language is to Ph.D. in Psychology, Anthropology, Humanities as correlation of 10,000 years of recorded history is to __________ of the universe.

The problem with physics, the problem with mechanics and the problem with contemporary computers is that they are stable. They don't mutate, or when they do you format the hard disk and reload the operating system. It is not considered good or evolutionarily beneficial, you don't suddenly get a more useful analysis of the effects of fatty acids on protein production of cells when your editor glitches during a file-save operation…you get trash. Instead of evolving, you start over, not the same thing. Imagine writing a term paper or love letter or grocery list, whatever. And to add to it, edit it, modify it, you have to rewrite the whole thing from the beginning. Our artifacts of technology are, for the most part, very brittle and not like our biology. The brain's synapses are understood to pass or inhibit a data stream. But that is an incomplete understanding. Neurobiologists know that the chemistry, the position and quantity of synapses, dendrites and axons make the brain much more an *analog thought ecology* rather than a digital device. The complex organ's function is not the same as the binary logic of a computer. The growing of a dendrite is not a photo reproductive process replicating to a low order of angstroms. Tremendous variability is allowed in the human brain. It is very difficult to design that potential for variation into computer hardware. Engineers have tried all kinds of things. Random number generators, fuzzy software, even neural-net software models and hardware. But we have nothing that produces as many flavors as the human brain. Or any other organic brain for that matter. I once thought that we could make smart toasters using bird cerebellums. When you wanted to teach it how to cook your bagel, you would trick it into thinking it was spring. Some birds learn a new mating song every spring in hopes of doing some fertilizing. We would make use of that. And the good thing would be that when it burned your bagel, you could call it a "bird brain toaster." You can't do that with an ordinary toaster.

I mention this to bring to the fore some measure of the difficulty of Melissa's task. The AOU was trying to invent a thinking being that was predictable out of organic resources. And he had to do this because where he came from it was the only way they knew of to build a universe. They couldn't yet seed a Force Bottle with mechano-forms. Or we would not be having this dialog as one-sided as it is. We would be mechanical dogs, or chimps. Probably chimps because cooks, waiters and bell-persons might need opposable thumbs.

A Meeting with the University President

I was about to backtrack again when the door to the lab opened. A sober salt-and-pepper haired woman entered, looked around and summed up the situation with an animal cunning appropriate to the bizarre nature of the day.

"I don't believe you people are authorized to utilize university resources. Please vacate this lab. Except for Melissa, whom I would like to visit her advisor at the earliest opportunity." She spoke with authority.

Of course that was just a beginning. Her quick analysis had not taken into account the cranium rings. Or if it did, she at first considered them ornamentation — maybe love beads. As she watched, the diamonds receded from view and both the AOU and Melissa returned to the living. They observed the intruder for a long moment; much longer than is considered politic when addressing the President of a major university.

The President can be forgiven for jumping to conclusions. We all walk the planet leading with our chins. We confront every new experience with our presuppositions. We confront even the most dangerous situations with a wall of impermeable material imprisoning the idea that we might not actually know what is going on. Our presuppositions impose a pitch black blindness and stripped-naked vulnerability. There we go, chin out, thinking we understand the opponent who is instead a wildly variable and unpredictable mutant butterfly with a nasty stinger, standing dukes up in front of us. As did the President. This lowly university President was standing toe to toe with the actual Author of University itself. She was confronted with the embodiment of magic and its cadre of enhanced human minions. I'd start bobbing and weaving Ms. President, or perhaps backpedaling might soften the coming blow. To my surprise Melissa, newly returned from another trip abroad, spoke first.

"Please, Ms. President, we are doing important work here. It will be better for you not to be involved. Go." Melissa would be considered a loose cannon in any social setting, any social interaction requiring the least political savvy. With her enhancements, think of her as a loose hydrogen bomb. Needless to say, the President didn't go.

The President, for her part, drew upon years of experience at handling all sorts of people with acuity and finesse. She went to the phone in the lab and began dialing Campus Security.

One problem with being 10,000 years old is that, for the last approximately 5,500 years, I have been of age for the military draft. Which means that for all of that time, any leader able to convince the people, subjects or followers to participate in armed conflict has called upon me to exhibit my patriotism and go to war. True, I have moved around a lot, but it is just a matter of time before I'm in the wrong place. You might now realize why I have become an ardent pacifist. Even if pacifism weren't the smartest approach to problem solving, I was sure to get a little tired of being asked to beat up or kill people. And yes, there have been times when I participated of my own choosing, but 5,500 years is a long time and most of those fights could have been avoided entirely. That is just a fact.

So, the idea of being confronted with Campus Security on the first day of my return to Princeton was most disappointing. Things had clearly gotten out of hand, and my physical changes, if they turned out to be positive ones, would be of value only in an atmosphere of relative freedom. But, as I watched the AOU and Melissa, I noticed that the skull ornaments were back, as were mine. They subcutaneously informed me that the phone in the lab would not be connected to Campus Security, and would I please answer the phone and inform the President that Campus Security was on its way? Apparently, the AOU's wall box could override some technologies when in close range. It caused a redirection of the phone line through a quantum junction box and broadcast it to my skull beads. Neat trick: who needs the Baby Bells? Say goodbye to another monopoly. I took the call.

We didn't want the President to become nervous. I had decided, for the moment at least, it was prudent to give Lexius my aid. To borrow from his western lexicon, he appeared to me to be the one with the mouthful of tobacco. It seemed like a good idea to prevent his being smacked. The AOU project plan showed some tasks still outstanding: he needed to stall for time. I would employ my proven skills in the tactics of delay and make use of the time, myself, to design an unobtrusive exit. So, the call was intercepted and patched into my brain. I assured the President that security was rushing forth and would soon darken the lab door and took over the verbal conversation.

"Melissa is sometimes abrupt, Ma'am, since you know her that can't be a surprise. Please allow me to suggest that you and the university might be well served by Melissa's unorthodox approach to her doctoral thesis." I was smooth as silk. And I was more than a little lucky to be wearing my professorship shell. As I indicated, we all lead with our presuppositions, and the President was no different. She didn't recognize me, which bothered her, but she was sure I was some kind of visiting professor, perhaps on campus to give a speech or award or something more unlikely…such as to learn something. She had clearly supplied a set of plausible scenarios from her own experience, because she adopted me as the person most likely to be influenced and asked.

"What is your involvement in this venture, Professor…."

"Marblemeister," I supplied. "I'm just visiting, and was caught up in the excitement of Melissa's work." It is better to speak only truth. When that can't be accomplished, attempt to speak no lies. It is much harder to maintain a fabric of lies than the truth. Have you ever noticed that? If you say no, I'd say you are lying, or measurably stupid, or perhaps just being obstinate.

"Since the work Melissa and Lexius are doing is of a temporal nature, perhaps I should do the explaining, if that is all right with you." I find that phrase curious. I'm glad it is generally understood. Because if things really had to be *all right* before making progress, progress would be slow, indeed.

"It is better to speak only truth. When that can't be accomplished, attempt to speak no lies."
—*Thodkin Marblemeister*

"How can I judge? I can only see that bumps are appearing and disappearing around your foreheads. Is that significant? If so, perhaps danger exists to which I should respond with flight instinct." The President had more than one way to say she was suddenly quite scared. And yes, there is a small incongruity regarding the analysis of whether to respond to one's more primal instinct. The instinct must already have ceased to exhibit the power of producing action without thought. Action before cognition.

"Please, allow an explanation. You have called security. They will certainly be here shortly. We will not endanger you unless you are afraid of ideas. Which is not to say we will endanger you if you *are* afraid of ideas. It is, rephrased, that you are in danger only of being confronted by new ideas. Should I continue reassurances, or may I go on with a presentation of salient facts, logic and physical proofs of principle?"

"No more assurances, please, they make me too confused. Feel free to begin whatever explanation you can accomplish before my campus resources of enforcement arrive to summarily imprison you in the campus jail cell. I must keep watch over University liability and I would appreciate knowing that nothing untoward in any form is taking place."

"You will be the judge; any assurances I could offer lack credibility anyway. First attend the projected images." I directed the President to the output from the electron microscopes being projected over the chalkboard covered with the scribblings of Melissa's unreadable shorthand. "They represent a breakthrough in two sciences and point the way toward world-changing technologies. I presume the University is interested in breakthroughs?" That was rhetorical, and the President didn't bother to respond. She just stayed as close to the door as she could manage while still being able to see the projection.

"Before you is an ongoing experiment. The person working with Melissa is attempting to discover why a formulation intended to produce non-mutating proteins failed to achieve its objectives. In other words, where it goofed up. You are seeing the first ever example of a working cell-level manipulator of genetic material. That is an achievement that could bring prestige to the University rivaling any achievement since *The Great Awakening.**

"This is one of two breakthroughs in evidence. Utilizing resources in the Nanostructures Lab, Melissa has migrated the protein manipulator to the atomic level, constructing the first atom pincer. As you might be aware, an atom pincer, its attendant quantum mathematics and a useful nanodesign software application is what has been needed to fulfill the Holy Grail of nanotechnology. Melissa alone understands how to create this tool. Atom pincers are the equivalent of a fire safe used by primitives to protect a burning ember during migrations. From this ember survival could grow. Before nanotech could blaze, somebody had to capture an atom pincer. It is the glow that will ignite science and a million new engineering applications."

• • • • • • • •

* A time of religious revival and sociopolitical change in the Colonies during which the town of Princeton won a small contest to be the new site of the secular college.

The President could not evaluate the supposed evidence. Still, something in the presentation created a sense of truth being told. It is true that truth was being told. But beware of relying on such a feeling when that is the only evidence which you are qualified to judge. The President, it may be said, was an expert at detecting bullshit. But just think what 10,000 years of telling stories might mean when developing bullshitting skills. Could it be that the President was beyond safe depth? Well, yes, it could mean that. But there is a benevolence irreducibly cast into this storyteller. And it helped that I was telling the truth.

I continued, "At the cell level, we require gene manipulation to accomplish certain science fantasies. Obviously, repairing or correcting genes will save many people who currently suffer disabilities or even death. DNA reconstruction techniques will extend life. If you think of the world playing itself out as a rear projection, wide screen movie, either of these technologies would slash a ragged portal allowing access to whatever is back stage. If you imagine each cell in your body is made from a picture puzzle of New York with 100,000 pieces, these technologies would make all the pieces square and the same size. Or you might think that living cells are a tall wooden ladder and each rung is a gene; and ribosomes surrounded by chemicals are the firefighters climbing the ladder to help some gene continue to express itself. These technologies are the hook and ladder on the big red truck with the spotted dog. These technologies are point-and-click compared to text terminals. They are carbon fibers versus the bronze age. They are Burt Rutan versus the Wright brothers; Martha Stewart's Kitchen versus Easy-Bake-Oven. So much is clear. I could go on with more obscure similes. As an example, Treplakin Eater-of-Bears versus Fodruv Lunch-of-Bears, a great warrior and a poorly coordinated member of an aboriginal tribe, but the obscurity might lack communicating effectiveness. Let me simply ask, would you agree that prestige will shower the parents of these achievements?"

As was predicted, the President's fear was squelched by the possibilities presented.

During my little show-and-tell, Melissa had completed her computations and so was paying attention to our conversation. The President explained the facts of life to us.

"We have specific procedures to follow. Melissa must present her findings to her advisor who is the chief investigator on most of our bio-research grants. I'm sure that, were he to approve the findings, Melissa would be allowed to continue her investigations."

"I would be allowed to be a member of the team investigating the work I have just completed? I would be allowed to give up my knowledge and all rights to it to the University?" Melissa chimed in without inflections of irony.

"Without argument, that is the way it works. Remember whose shoulders you are standing on," replied the President, attempting to adopt authority by proxy — she hoped to channel people Melissa might actually respect.

"You refer to my professors and research principals? Whose shoulders are they standing on? Who is acknowledging Feynman or Drexler? Who pays royalties to Einstein's estate? Which opportunists?" Melissa was probably not helping, but it mattered little. I was certain the conversation was futile at the core.

Millions of Micro-political Ecologies

Universities are political institutions. They behave as microgovernments, enforcing specific beliefs, just as religions do. Politics is religion. Religions are bureaucracies. Bureaucracies benefit the bureaucracy first, the politician second and if there is something left, it is used to provide motivation for the body-politic. Motivation to promote the buttered side of the bread. The buttered side of the afterlife. The buttered side of the pork barrel. There are many well meaning members in politics, but the trend is clear over recorded history. Even the dictionary has been allowed to notice the essence of politics.

As yet, it has not been discovered how to provide for the well-being of the *many*, without politics. Politics is a naturally occurring Darwinian strong attractor in the solution space of governance. And the many examples make it even more difficult to invent alternatives.

Main Entry: **re•li•gion**
Pronunciation: ri-'li-j&n
Function: noun
Etymology: Middle English religioun, from Latin religion-, religio supernatural constraint, sanction, religious practice, perhaps from religare to restrain, tie back - more at RELY
Date: 13th century
1 a : the state of a religious <a nun in her 20th year of religion> b (1) : the service and worship of God or the supernatural (2) : commitment or devotion to religious faith or observance
2 : a personal set or institutionalized system of religious attitudes, beliefs, and practices
3 archaic : scrupulous conformity: conscientiousnes
4 : a cause, principle, or system of beliefs held to with ardor and faith

Main Entry: **pol•i•tics**
Pronunciation: 'pä-l&-"tiks
Function: noun plural but singular or plural in construction
Etymology: Greek politika, from neuter plural of politikos political
Date: circa 1529
1 a : the art or science of government **b** : the art or science concerned with guiding or influencing governmental policy **c :** the art or science concerned with winning and holding control over a government

But, if you look deeply enough into the ramifications of Melissa's potential, you will see the doom of traditional politics. It would not surprise me if the President of the University was already cognizant of these potentials. Great politicians are both learned and intuitive. They perceive innately some things that others must experience. If they are successful in politics, it is largely because of the innate ability to discern opportunity and danger for the machine enforcing their beliefs. Think of governments, religions and their ilk as huge stone wheels, a mile high and half-a-mile wide. Once in motion they roll over First Minds which either flatten like little thin pancakes or get stuck in the stone and are carried along. It is a lot better to step aside, if you can, than be flattened or converted, because options diminish quickly, lose depth, become two-dimensional. The stone turns, but is very slow to alter direction. The rolling bureaucracy gathers no boss. Well, that is so imprecise as to be untrue. But, it had a nice ring to it.

In fact, it is possible to appear to be the boss. Penetrate the University President's inner thoughts and you will discover a belief that in spite of the board of regents, advisors, legal restrictions, government regulations, and financial purse strings, she thinks she is the boss. Still, it is the stone's inertia that is the real boss. Look at the huge granite stone from the side and you will see a bunch of people on the crest madly running backwards as the stone turns under them. They provide

Main Entry: **pol·i·tic**
Pronunciation: 'pä-l&-"tik
Function: adjective
Etymology: Middle English politik, from Middle French politique, from Latin politicus, from Greek politikos, from polites citizen - more at POLICE
Date: 15th century
1 : POLITICAL
2 : characterized by shrewdness in managing, contriving, or dealing
3 : sagacious in promoting a policy
4 : shrewdly tactful
synonym see EXPEDIENT, SUAV.

"Politics is a naturally occurring Darwinian Strong Attractor (DSA) in the solution space of governance."

"What the hell?"

—*Thodkin Marblemeister*

"I dreamed I dated an exciting, beautiful, provocative, mysterious and challenging future."

"I said it was a dream."

—*Thodkin Marblemeister*

"Most issues confronting humanity are just leftovers in the refrigerator of history. They are just problems that have only been partially digested, while the remainder rots.

—*Thodkin Marblemeister*

the energy that helps the stone overcome entropic forces. Believers and dupes make the little additions that cancel the entropic subtractions of change. Or not, leading to a slow extinction of the religion/government/school/business as it trails squashed, ineffective little personalities stretched too thin to provide sustenance for feeding the enterprises of life.

It is not impossible, but it is difficult to develop latent political skill. It is an objective for the well developed personality, the optimal First Mind, to be tactful, sagacious and shrewd. If you leave out the sinister from the definition of politic, you have an evolutionary quality of personality. In contemporary personalities of the President's level, it is common to observe contradictions. The worst politicians execute political agendas supported on unfounded belief systems. It was up to me to discover what we were dealing with in our current involvement.

I Had a Dream

But now I am reminded of a dream I had once. Yes, just once. I was at a girl's house to pick her up for a date. Some members of her family were at home so it took a little while to negotiate the pleasantries. Then, just as we were about to leave, her father returned home from a political rally. A handful of people entered with him. They were all dressed nicely in conservative clothes. They wore bow ties and evening gowns. Not everyone wore both.

The women wore evening gowns. The men were in suits and everybody had black overcoats. They came in, very quickly moved chairs from around the dining room table into the living room, and began talking politics. And the girl's father poured everybody glasses of Scotch on-the-rocks. I don't know why Scotch was important. Maybe the problems would too easily reveal obvious solutions without the obstruction of muddling intoxication. Maybe that was symbology for a lack of mentalic training. Anyway, I was just trying to get out of the house to enjoy pursuit of worthy conquest while they wanted to know what I thought about one trumped-up issue or another. I remember one issue had to do with mashed potatoes. But the issues were really the same kind of little issues that always occupy people when they become involved in politics.

And that is a problem in the First Mind problem space, because these issues were much easier and obvious before they became political objectives. If we could just think better, as a species, if we could reason better, so many of these life-wasting childlike initiatives would just go away. And then we could all go out on a date.

And so I Asked Another Question

Where are we in this story? Let's review.

I have met the Author of the Universe. His name at the moment is Lexius Acura. He took over the body of the former Lexius just after he died. He is here on earth to discover where his invention went wrong. Really, where his cookbook went wrong. He has co-opted a scientist and myself to help him calculate because he doesn't want the tabloids back home to know he screwed up. He has made unauthorized use of University facilities and resources. And he now confronts the President of the University who stumbled upon the skunk works in one of those improbable happenings that take place only in story books. Or in real life just when you think everything is under control. The President has called the campus cops and is just being made aware of huge potential profit for her political interests, and is trying to assimilate it all while developing rationalizations and tactics to close the rest of us out.

Little does she know that the campus cops aren't really coming.

Campus Security Arrives Anyway

Sooner or later, but not before....

At nearly the same moment, the President and I notice a bubbling vat right out of a cheap sci-fi thriller. The vat, situated casually in a corner behind the AOU, had seemed an unused artifact in the lab, but had actually been filling so slowly with the green goop that it had provoked no noticement. It wasn't very deep, maybe 9 or 10 inches, and about 24 inches in diameter. An agitated green liquid roiled an inch below the top of the container. As we watched, fibrous metallic tendrils formed at the edge of the vat and grew up the sides clasping mechanical fingers to the rim. It continued to grow until, very quickly a sort of arched truss was created and a flexible wire extruded from the apex. A shiny disk like a coin formed at the end of the thread. It looked like somebody had dangled a silver dollar from the truss. The coin was balanced perfectly parallel to the plane of the fluid below it and lowered until it just touched the surface of the liquid then all the growing of metal stopped.

I allowed my attention to take note of the President. Had she been Caucasian she would have been white as a ghost, or blue from not breathing for so long. As it was, she just lost some richness of color. And a little perspiration dotted her brow.

"What is that?" she articulated.

The AOU replied. "That is the outcome of Melissa's work. It is a test run of my experiment. You might say it is a prototyping tool."

"And the green liquid?"

"Think of that as primordial soup. It is a mixture of nutrients, minerals with some special attributes, some basic molecules and something rather new. Something like nano-T-cells. You see, the first time I relied on gravity and tidal forces to produce replicator molecules. Little combinations of compounds that managed to recreate copies of themselves. That was really the foundation of life."

“Yes, I understand the theory. Once the oceans became liquid water they pushed simple molecules into traps like coastline perturbations where they collected, being occasionally bombarded by cosmic rays and joined by other enhancing materials until more complex structures mutated into existence. God had nothing to do with it, I understand.”

“Do you? Well, just add five billion years and you get some very complex replicator ecologies. Like animals and plants and bacteria and, well, you understand all that. The problem is that chance played way too big a role and to my sadness produced intelligent replicators, self-aware intelligent replicators. I don’t want that to happen again. So Melissa has taken a lesson from my mistake and a goal of your budding nanoscience and designed a device that is something like an erector set at the atomic and molecular levels. She has designed composites that will *tend* to come together in desired configurations when quantum organizing forces are applied to receptors embedded in the Force Bottle barrier — its stasis boundaries. Like growing crystals, only the seed will not have to be in the universe, and these quantum pattern tendencies will ultimately produce little machines. And those little machines, rather than evolution, will then construct only specified replicator molecules and those will create only organics of use to my goals. And further, these little machines will disassemble anything that is not part of the program.”

“Who are you again? And how is it that you feel justified in using my University to attain your goals? And your goals are?”

“Vacation, and I don’t need no stinking justifications.” The AOU slipped into a drawl derived from Texas and Mexico — think of it as his not-of-this-world interpretation of the classic scene. Discerning observers have by now been bludgeoned with the evidence that he was an admirer of the western. None would guess he could recite any line from any western ever made. “It is true, I can’t shut down your universe, but there is nothing that says I can’t use it for whatever I want. And if you weren’t such a purty lady, I’d advise you that when you’re throwin’ your weight around, be ready to have it thrown around by somebody else.”

I felt the conversation was about to take a nonproductive turn into the Twilight Zone, so I stepped back into the moment. “I don’t believe, Ms. President, that you have been introduced to the Lexius Acura.” I was wondering if I could avoid convincing her of his being AKA The Author of the Universe. “Lexius has offered Melissa and me some proprietary information with which to leapfrog certain boundaries that confront contemporary science. Melissa is attempting to verify the information with the little experiment in the vat.” I had now reviewed the details of the green goop.

“Please inform me further of the experiment. I may need it to be stopped.” The President, confused by the “purty lady” remark, had ignored Lexius’s indignation. “And don’t think I didn’t notice that this gentleman claims a name created from two competing makes of Japanese automobiles. Were I not already certain this is an illegal undertaking, I would have been made so by the name alone.”

“Actually, I think that Lexius — whose name is both unusual and unexplained — was attempting to make the point that when confronted by his priorities the law is a lower court. He has power a la *Treasure of the Sierra Madre*. He has us out gunned. We are a child’s popgun, shooting cork projectiles. He is an asteroid flinging force, deleting the dinosaurs. Whoever he is, asking permission is not one of his known behaviors, as the beads that appear around my forehead would attest were you fully apprised of our circumstance.”

I found myself mimicking the President in tone and verbosity. I could have just told her that she was spitting into the tornado.

I would have reasserted ownership of the explanation, but a picture is worth a thousand confusions. As we watched, the little silver coin was smoothly ascending. Progress was slow but discernible and from the bottom of the coin a green — don’t ask me why everything is green — a green pair of oval shaped feet and what appeared to be ankles were forming. While it looked like a creature was being raised from the fluid, that is not technically what was happening. The feet, ankles, legs and what would follow were actually growing — atom added to atom to molecule to molecule. Little pods had formed near each…leg? They looked like small soccer balls with a tube extruded from one end. They pumped and I could see minute expansions of the tubes in a rhythmic cycle. All the while the feet raised, the knees formed, et cetera. The President was frozen. I was somewhat taken aback even though I had a detailed specification of the process in my mind. This was eerie. Not the science. But, just think. Two days before, nobody on the planet had a clue of how to build anything practical with nanotechnology. Now, something obviously complex was forming. I began to think this was just a rubber toy, or some neat trick like investment casting.

By now, we could see something very similar to the construction of a human, up to the shoulders. The neck was a little different, somewhat more wrinkled. But, minus any imperfections and any special character, the thing was identifiable as a bipedal animal.

"Oh, my God!" The President found her voice. "You'll kill it! Get its head out of there!"

Well, she hadn't quite understood or had become seriously overloaded with anxious, conflicting inner voices. She obviously thought they were doing some kind of Houdini with a newborn or something. But, Melissa attempted to calm her.

"No, it is okay. There is no pain. That is why we are making the brain last. The neurons are designed with a start button, nothing happens until we are ready…no pain. There will be some confusion until they are enabled by the janitorial nanomechs, but I assure you there will be no pain."

"What is it? Is it dangerous? How could you be making a thing like that without any precautions. You are all going to jail..

Alas, that might have been a true statement. This could have been risky. But, Melissa was very smart, roughly equivalent to every scientific genius in our history combined. And I had a new-found ability to do a fairly rigorous validation of her work, not to mention the AOU's use of intelligence beyond comprehension. In about an hour the nanomechs would disassemble the green thing and then themselves. And it wasn't made like us. We didn't share any nexus point in evolution, and we wouldn't interbreed. It wasn't toxic or allergenic. It was benign to organics. The creature was not going to be self-aware in our traditional way of thinking. It would be more like a very smart dog without the craving for being fed treats at the supper table. And it would already know how to do a task very well. What's more, it wouldn't care about things it wasn't designed to care about. It would be like a Republican. The vat represented a model of the Force Bottle initialization only instead of taking several billion years, it was taking about half an hour. Pretty good for a species that was nothing more than accidents of mutation.

And there it was. A cute little green thing. The AOU went over to the still-hanging creature, dislodged it from the shiny disk — like picking a grape from the vine. He turned it over and from his pocket he pulled a small object. It was, in fact, a pink hair comb from a Barbie dress-up kit. He held it out to the creature. Its little eyes blinked. It reached for the comb and, clinging by its now long padded toes, began to comb the AOU's hair. Stroke after stroke, gentle with unkempt knots, it combed it straight. The AOU's eyes closed.

As sometimes happens, chance takes over where plans leave off. No one in that lab had words for the event except the AOU, who wasn't speaking. The rest of us were still essentially human. We simply had no experience allowing us to relate to what we were seeing. I can say that in 10,000 years I had seen nothing like this. The enormity of the achievement and the Byzantine expression of the proof was just un-reasonable (sic). He might as well have made a tail-wagging Puppy Vacuum. So we were standing quietly, even the University President, watching a green, intelligent construct comb the hair of an entity who had just certified its claim to be the maker of the universe...when the Campus Constabulary entered the lab on its appointed rounds.

The President, with the least to evaluate, snapped out of it first and directed the Campus Police to arrest us and call HAZMAT immediately.

Sometimes Quick Action Just Isn't Enough to Save the Day

"Melissa, you will be incarcerated. Please tell no one what you have done. I must consult various...well, consultants." She meant that she needed time to figure out what could be made of this opportunity. She was going to shut this up and deal with it at her leisure.

Melissa's forehead grew bumps. When told what to do she most often produced inverted results. She was about to take action. She may not have been a political animal, but she did understand every language on the planet now. And the word *incarceration* or its equivalent was in all of them. So, I too, popped some beads. I was beginning to have some volitional control over my new self. We communicated. I requested that she withhold action until we could extricate ourselves less obtrusively. I had contingency planning of my own freshly minted.

The AOU was busy planning too, although it might not have seemed so. For nearly 15 minutes the green thing had been combing Lexius's white hair. It was now very straight and had a lustrous shine to it. Eyes closed, a gentle smile on his face, Lexius fell from his stool and died. The Comber rode him to the floor then diligently made adjustments to accommodate the AOU's inelegant sprawling. When it was as neat as it could be made with Barbie's comb, the creature searched the room for someone willing to be combed. It spied the President and something in it decided her salt-and-pepper cornrows represented a proper challenge…she could use a new do. With a gentle yet surprisingly quick grace, the green thing attempted to shinny up the President's leg. Around the lab the two of them danced. The Green Thing showed only two oscillating emotions represented in amazing similarity to human facial expressions. The first emotion, represented by an upward puckering of its denuded brow accompanied by a subtle backward movement of the little green head, was an exasperation at being thwarted in its simple goal to create the perfect coif. The second was a cute little frown of revitalized determination to do what was obviously needed, even when confronted by hair styling ignorance. The President alternated between screams, curses, commands and yelling at the Campus Police to shoot it. To their credit, they didn't shoot her.

Melissa and I Take an Opportunity to Slip Away

In the chaos, Melissa and I slipped away. Closing the door behind us,

I just happened to notice the jukebox selector flare and evaporate. The remaining substance in the vat did the same. Presumably, the green creature would feel no pain as it disassembled. We couldn't wait to see; we had no time to participate as observers. It was time to leave town. Sometimes one day is a proper length of stay. I was leaving way before the fish would start to smell.

We Leave Town

It was late evening and we just missed the Salvation Army store's closing. So we were forced to go to the mall. We needed to pick up a few things. First, we needed hats. Melissa would be convinced to supplement the Baltimore Orioles cap. Something with a brim would be good. I wasn't sure of Melissa's fashion sense. But I had always liked hats. Paying full price went a little against the grain, but this was an emergency. It was time to take a train, spend a few bucks. And maybe I didn't mention it but I have a few bucks. In fact I have saved up. Compounding interest is a marvelous thing. When you are frugal and live as long as I have, interest really mounts up. And I have learned to be inconspicuous about it. You never know when buying yourself out of a jam is the right thing to do.

We got hats. We got clothes. We got luggage. We ate. Then we hopped a train to Boston, well New York, then we would go to Boston. I had a stop to make in Boston. Melissa, only too well, recognized her limits of worldliness and took my lead without comment. Truly, she didn't want to be bothered. I guess the train to Catatonia was just one track over.

You might think that we now knew literally everything, But, the AOU gave us only what it was interested in. We had language, science, some miscellaneous research, but even though we had great available capacity, we didn't have many immediately useful things. Like train schedules. We needed to shore up our practical knowledge.

I Review some History Regarding Fluidic Consumable

One thing we would never be ignorant of was beer. The AOU had a marvelous database of beers, complete with his own critique, deduced recipe and chemical analysis. He also rated each beer. The highest ranked at the time of his demise was an obscure homebrew named Bittah Blue, which had a quaint story attached to the dossier. If we have time, someday, maybe I'll relate it to you. One of the ingredients the AOU had cataloged in Bittah Blue was common Northwestern Bathtub Mold. It was to the chemical contribution of this element that the AOU attributed the quality. Once again, we see that chaos plays a role in everything...small things can have big effects. Apparently, no other beer the AOU ever drank included just the right amount of this ingredient. I dare say no brewmeister will ever quite equal Bittah Blue, which is a shame, since the AOU consumed the last known bottle of it preserved some 20 years in a stasis field. During one of his previous incarnations, the AOU had crashed a wedding reception, sampled an assortment of home-brewed refreshments, and preserved the last bottle of Bittah Blue with the help of a "Porta-Stay-Forever"* disguised as a $1.99 Styrofoam picnic ice chest. That was the day he hatched the plan to return and figure a fix for his next Force Bottle. If mere humans could create Bittah Blue, surely they could help with something just as important. And yes, to give this section a title of its own was probably a little flagrant. Would you feel better to know that I have my reasons? Good.

There were other references in my newly loaded memory that were

• • • • • • • •

* It seemed I also had the basics of "Porta-Stay-Forevers" in my new library. I found it counterintuitive. I have always thought that stasis fields would stop all atomic activity. Obviously, that is impossible. But it turns out that it is possible to take the equivalent of a quantum Polaroid. When you push the freeze button on a PSF, it takes a sample of the rate of atomic decay of the object in question. Then, begins a programmed loop of universe samplings where the decay is missing. Then, using a quantum transmitter, it replaces the object in the field with the newer version, which decays, is replaced and so on and so on. Put simply, it uses a quantum-refresh storage technique. I should have thought of that before.

curious. I found a use for a citation regarding frogs. Certain frogs have evolved a protection that confuses predators. Stripes running down the frogs back obfuscate the signature shape of the frog. The frog is not recognized in the backdrop of the flora. This is what Melissa and I were doing. Wearing shells that might be known as "Investment Banker," we boarded a train to New York. Melissa required some coaching in order to achieve this look. She was typically without a shell. She was deeply "Geek Grad Student." No shell, just meat. I wonder what it would be like to be young, really just 20 years old in this time.

I remember seeing a kid forced into a ring-bearer's tuxedo. It took about an hour. Lots of fidgeting, scratching, fussing. Then the mirror. After the wedding, they couldn't get the kid out of the thing. Melissa had a similar experience. My hope was that she would not become a primping teeth-whitener and all that entails. I might have to do an intervention.

People wouldn't be looking for "Investment Banker." They would be looking for "Professor" and "Geek." And if you want to blend in on Wall Street, maybe "Investment Banker" wasn't best, maybe "Broker from a Big House" would be better. Still "Investment Banker" was good. We even had a task to sell stock in New York, before going on to Boston. It was time to make a change or two.

Never Judge a Book by Its Cover

Now, you may have noticed that a book subtitled *Conversations with the Author of the Universe* was in jeopardy of committing unorthodoxy by killing off the Author of the Universe in the first 60 or so pages, hence making conversation more difficult. And I agree with you if you were thinking such. So, even though we are going to keep the subtitle *The True Future* for now, the section name must be changed. The new name is…

The True Future, a Predicted Evolution

That seems pompous enough to replace *Conversations*, etc. We'll go with that.

The fact is that the AOU was still with us, sort of. He had left us all his research and a codification of his personality, should we figure a way to create some pretty sophisticated AI hardware. Or should we figure out his trick of taking over dead people. But, I was thinking we could do without him. He was more than a little self-centered and things had a way of getting out of hand when he was around. I say, "Let sleeping Authors lie."

I must have said or broadcast this out loud, because Melissa replied, "I agree. I don't like being told what to do."

This brought us to a sensitive topic. "Melissa," I delicately asked, "are you aware of any limits to your currently existing abilities to socialize without shooting your objectives full of cannon-ball sized holes?" She stopped to review the question. It took a while. She is very thorough and she has lots going on inside that the rest of us would probably think was minutiae, not worth considering. I have found when dealing with super-intellects, it is best to let them deal with minutiae. It seems to be one capability that separates the men from the boys…so to speak. Quick surface thinking is very easy to do. You get an impression, it makes some sense and you just can't be bothered, or just don't have the tools to consider it further, so you act. When really smart people do that, the results are pretty good. But what about when really smart people deal with all the little chess moves, stretching out into the past and future? That's when we notice genius. There is a miscomprehension about what it takes for a genius to have a good idea. People think it just blinks into view. Well, that part is true. But more often than not what blinks into view is wrong. And after some serious grinding of mental cogs, some torquing of whirring shafts and stuff, there is nothing left but a dead end, wrong idea, stripped gears, bent rods, gears with no cognition. Then it all happens again in the next idea. Melissa is all of that. She is genius of all eras incarnate. It takes her awhile to explore anything she really cares about. I take it as a compliment that she considered my question any time at all.

"Your analogy appears to be warranted. Teach me to be a better politician. Teach me to make smaller holes." Short and direct, that is

Melissa. I liked the evidence that she understood perfection was an illusion and unattainable. She would be successful if she could make progressively smaller and smaller holes in her objectives. Complexity prevents perfection. Perfection implies a single solution. To think something is perfect is to make it obsolete and then uninteresting followed immediately by trivial. And we did have a little travel time. With our nice new hats, we should be able to have a surreptitious political training session. So we gave the new beads a little runup.

Politics Second Hand

We got the last train to Clarksville. Actually, that is a song, and it may have been available from the AOU's jukebox, but we really got the Amtrak to New York. We managed to secure two seats where we could be unobtrusive, but having head beads was a convenience. No one would be overhearing any conversation we had unless we wanted them to. The beads were a revolution, but it was clear there were some mundane enhancements we should have, like being connected to the Internet and telephone networks, not to mention satellites. I would solve that in Boston and other stops I was planning to make. For now the beads were plenty powerful and surprisingly friendly.

"I would like to become more effectively social," Melissa began. "But first perhaps you could tell me what you have in mind more generally. Upon quick review I recently noticed, during our very silent dinner, that I have known you for less than a day. Six point two five hours and some left-over, but I don't want to become a Spock-a-like, so I won't be stating everything to excessive precision in the future. It is unusual for me to blindly follow someone I don't know. And unheard of to blindly follow anyone I do know. It is only the technology that we share that made it seem like the right choice at the moment. That and the fact that I have been just completely enthralled with the changes I have been undergoing, not to mention the task I was put to. But I am taking a moment to deal with short-range plans. So, as we sometimes say…waz'up?"

This was more dialog than I had ever heard from Melissa. It gave me hope that she was going to survive the keys to the kingdom still somewhat recognizable as herself. And also bear in mind that she said it all without moving her lips and at a tremendous data rate. The data rate at least is significant. Humans are clocked at about 1.5 thoughts per second if you consider recent studies that some multi-tasking takes place (of course, nobody who reads this is that slow). A lot of processing went into that thought and a half. A single neuron fires in about 5 milliseconds. But there are 100 billion neurons each with about 1000 connections to other neurons. So it is lucky it is all a network or thinking would take much longer—maybe 15 years . No matter how fast the brain is the speech would have taken much longer before our enhancements. This notified me that we had been changed in some unexpected ways. I communicated this as an aside to Melissa and she provided some details.

"Everything is Ecological. We just might not be able to live in one or another."

—*Thodkin Marblemeister*

"Yes, we are not using our brains for much more than personality and goal setting. To support our nanobiological computational and memory enhancements there is a predictive algorithm that has never been seen on Earth. You know there are academics who spend their whole lives just thinking about the most efficient and clever ways to do some very tiny but important tasks. Like the drawing routines in your computer. How do images get clipped just at the border of a window? There are easy but slow ways and there are clever ways with great speed. The clever ways have been invented by some very myopic investigators. They get algorithms named after them. But in our heads now there is a complex process running that is attempting to predict what we are likely to want to do. Imagine that. We are running a predictive engine that is trying to figure out what we might want to do."

"And it doesn't just wait for direction?" I asked.

"No, that would mean it sat around doing nothing when it could be exploring possible futures like a chess grand master. It is already doing the job before we decide we want to do it. It is a work-around really. It allows us to keep the biology in our heads without many of the limitations. And the predictions will be getting better. In fact, we have just barely begun to learn about ourselves, but the predictions are already better than any profiling methodology in existence...by many orders of magnitude. Say, at the moment a thousand times more accurate. And this yields many more choices than we would previously consider given the same time to do the considering."

"I agree this is exactly true. I have noticed much clearer thinking. I am sure what is our most optimum course of action now, given my knowledge of you — somewhat limited — and my knowledge of human behavior — pretty solid — and my understanding of possibilities within the problem/solution space of my life motivations. Let me add that although I am directing our current flight, I do not assume that our goals are aligned. I am hoping that a sharing of secrets and life force drivers will result in a consensus. But, I am not presuming. I will say I am intrigued in a nearly overwhelming way for the first time in a very long, long time. It is not easy to upset my apple cart. I am very difficult to sneak up on. I have lived, coped and learned through many unforeseen circumstances. It is now rare to be surrounded with so much uncertainty. Today, I feel nearly young again."

A burst from Melissa interjects: "I suggest that is partly due to physical changes in your body. But I know what you mean. I was modified only a few days ago, when the AOU decided to forward his plan of creating a test case. Still, even in that few days, my body has undergone several significant changes. My muscles have gained tone not normally attributed to a geek. I'm a little heavier, but leaner by a great deal. My boyfriend of detailed intimacy, were I to have one, would not recognize my shape across the quad. I am metabolizing food more efficiently. My chronic head itching, something I have always thought was gross, is gone. I have desires for the same old foods, but find myself choosing, even liking, better foods as well. My sleep time is shorter, but more restful. I seem to be experiencing daytime dreaming. I haven't understood all of what has happened, but I am breaking it down. The AOU jerked us around pretty good. He certainly fucked me. If I wasn't so thrilled by the child he grew in me, I'd hate him for his rape of my genes. As it is, I think he is an asshole for his lack of disclosing the imminent by-products of his 'little-tuning-up.'"

The child statement was a metaphor. I didn't have to ask for clarification, because it was communicated with an attribute marking it as a metaphor. We were communicating nuance attributes attached to the content. I found the possibilities disconcerting. It had an upside and a downside. Or maybe I just wasn't thinking of all the possibilities. I first thought that sending a billboard along with every joke, metaphor or ironical statement would be less interesting. But, maybe I could only send the attributes if I chose to. I'd have to check this out. Either way, it would change storytelling forever. Miscommunication and partial communication have their place. They produce the unexpected. We need that for evolution. There was much to explore as the sentient prototype of an upgraded species. Melissa and I were beta sites. There were bound to be some bugs...which we were going to have to cope with using our own resources. The company that adjusted our wetware had abruptly declared bankruptcy and been dissolved.

"So, to answer your question, let me first reveal the AOU indicated to me that in order for humanity to survive the developments you were capable of bringing about, it would have to have personality training courses. I think of it as giving the keys to the family car to a trike-riding two-year-old. Much automatic, out-of-consciousness judgment must be learned between the trike and the auto. He picked me...he was going to pick someone, but stumbled on me. When he discovered I was more experienced than most...well anyone, he charged me with the task of solving the personality development problem and washed his hands of it. I have been thinking that he may not have been too bright about some topics. He didn't live as a human, he just invaded. He was created in a wholly alien culture. To the extent that sentient species must solve certain common problems he could relate to us. But clearly he didn't care too much about us. He had a tell-not-teach behavior. He told us what to do, used us for a bit and wandered off.

"To be candid, I'm glad he is gone. But also, I want to do what he suggested needed doing. I would be pleased by two things. I want you to express yourself...no task for me there. And I would be thrilled if the species were to survive and evolve...which, given some time and resources, maybe I can help. I really don't want to be the last man standing...of the few things I dread that is right up there with never eating another helping of world-class potato salad."

We continued to dialog, and as we did, I found my thoughts wandering. It wasn't that I was paying less attention to the topic we were addressing, but I just had more attention in the bank. I was suddenly able to pay more attention. I had more attention to pay. So I was able to notice the dreamlike state she had mentioned.

Life, Light and Darkness and Light

I opened a door to a nearly pitch black room and stepped in. I floated; the room drifted in space. The walls were covered with light switches of all types, yet neither the floor nor the ceiling was close enough to see. I sensed a cold, hard darkness trying to manifest itself in the room. It sought to take the form of all space. I flipped on a nearby switch and darkness shrank. A dim and soft light filled the volume near me. I could see the switches on the walls because of a quiet radiance that was given off by them. But the power source to any switch soon diminished and darkness tried to grow again. It was a subtle, purely uncolored shape, barely discernible from the ambient dark that owned the room. I turned a dining-room-romance-enhancing-chandelier dial fader up to maximum brightness, the darkness shirked its bully shoulders and camouflaged, but the light soon wandered away, a retreating glow of flashlight and dying batteries. I found a row of toggle switches, flipped as many as my fingers could comprehend and the darkness took a step back from where it crept...receded and patient it waited. This was the story of my life. I must find another light, and another and another...the darkness is never defeated, it is patient. Still, it need not win.

Concurrent Discussion and Experience Transfer

While I was having daydreams, the business at hand was ongoing. We were attempting something never before achieved.

It is a desire of humanity that offspring learn from the experience of the progenitor. Folly, of course. First and foremost, the environment in which knowledge is obtained is evolutionary or at least transitional. No two generations have ever been equivalent. But a common force in each matured generation is to make the world stable for the next generation. To make it what the incumbents know is right. We would be living in caves to this day...not such a bad place to grow up as I remember, but we'd have only the one choice if the incumbent generation had its way. So, I wasn't trying to give Melissa my deductions or my encapsulation of how life should be lived by showing her some of the details of my long existence. Rather, I had compiled a series of annotated experiences and consequences that I was hoping would permit her to rapidly create self-generated working hypotheses to be used as predictors of the potential results from her social interactions. It might enhance her use of words, her choice of metaphor, or a use of metaphor instead of declarative dialog. And I was interested in challenging her certainty. Youth lives in a paradox of certainty and uncertainty. In some things youth is fearful because, deep down, young people know they have no experience. But about other things, they know they hold truths, truths missed by their elders. The elders ignore these naïve truths because life has demonstrated its complexity and unpredictability to them. Youth ignores the elders' admonitions because it knows what it feels like to be right. We feel rightness when the mind is inactive after a decision point has been reached. We know rightness by its lack of activity. Neural networks perform an empirical calculation. If the results of a thought (a series of resolving attractors) comes to rest, causing only the smallest ripples, we interpret it to be right. The learning of youth is fast, but it is also short. So, for any conjecture, any extrapolation based upon experience, a youthful mind relaxes into certainty quickly. It is mainly the doubt produced by the complexity of experience that permits the sustained agitation of uncertainty. And it is uncertainty that allows for change, for options and for choosing.

So Melissa was absorbing, in a way that no human ever had before, the details and nuance of another person's experience. She was comparing her desired behavior against a richly presented library of perceptions from another life. It wasn't the same as living it. It was only a small percentage as real as life, but it covered 10,000 years, so patterns could be perceived and separated from the noise of my own biases and personality.

"How may I ever come close to thanking you for this?" Melissa finally spoke. "Somewhere in this course I realized just how personal this presentation, this offering was. I understand now something of the level of the force you are harboring to help our species. And I can't say I would have believed it communicated any other way. The AOU told me he was amazed you had lived so long. I didn't believe you had. I was sure you were just another nut case wandering the planet. I know a lot about biology. I couldn't fathom how yours could be so regenerative, but that is just physics and chemistry; I have always thought just about anything was possible in physics and chemistry. What I was sure of was that no one could live so long because no one would want to.

"And now, it seems I, too, am to be tested by such longevity. At this moment I find it inconceivable that I will ever match your achievement. This sharing of yours has changed me. In just forty-seven and a half minutes, give or take, I have realized that just doing, just finding my limits is not enough. I must have more. I must have direction. I must have something roughly equal to my potential to balance my choices. I need something big enough to get me up in the morning."

"Is there more that you can explain about that?" I needed her to arrive at a foundation for her near-term future. She had to go farther than the word "something."

"Yes, I see you are right. I must decide. It is clear that I have never imagined enough. Before my enhancements, I was pretty bright."

"The AOU called you a closet genius."

"Maybe. I was able to do things most of my colleagues can't. But now I can do things nobody can do, except maybe you, and I have no ideas from last week that are even a little bit in question. I can do anything I have ever thought of doing. If I don't watch out, I will be like one of your children. I will self-expire because I am unchangeable."

"Each of us is different. My children weren't you and you are not they. Your personality has contained a kernel that it expressed every way it had available. You are aggressive. You are carnivorous. And your favorite meat is invention. You now have tools you never imagined. And the forces for invention, for expression, are still there. In fact, they are probably temporarily paralyzed by ecstasy at what has happened to you. There is a huge difference between potential for achievement and achievement. And there are many lessons that have nothing to do with your ability to do science and little to do with your ability to invent, that will have to be learned — first hand. You will be tested by essential mundanity. Can you learn to keep you specialness from inhibiting your ability to enjoy your cohorts? Have you ever enjoyed cohorts?"

She agreed. "Yes, and there is a problem of scale. I have always worked by myself, taken my own counsel. And I have worked on manageable problems to a scale of my choosing. I have no comfort about taking on world-shaking endeavor."

"I have learned that it is better to be a world shifter than a world shaker. You are so young that you might think everything must be done today. Firstly, it can't be done today. Secondly, that is good; the world isn't ready to be shaken, and never has been ready. Kublai and Genghis Khan, Napoleon, Hitler, Triple Bear all shook the world; where are they now? Ideas shift the world, followed by other ideas that shift it again. Mine now becomes the task of trying to make it ready. We need new ideas and ways of proving them. If I may offer a suggestion. Adopt, temporarily, a sharing of my little quest. Perhaps preparing to reinvent the world, and doing it with a counter-Bondian 'stirred but not shaken' is big enough to get you up in the morning. I will help you get started, not that you need much help once you have set your mind to something. But, there is still only one of you. You are in danger of becoming the nail that stands out. This will not forward your plans nor extend your life. All of my experience is demanding caution of me too. It is time to fade into the scenery. It is time to become more than one wandering storyteller for a while."

"You are obviously working to some plan. What can you share of it?" I had her interest. She was recentering a little from the flurry of her last several days. She was beginning to exhibit some survival responses.

"I have lived frugally. From time to time I have made money, investments," I began. "It would not be an overstatement to say that I could discreetly fund something akin to a little specialized think tank. It would not be an overstatement to say that I could fund MIT. Which has led me to the following tentative plan…."

"Hah!" She laughed at my admission of wealth…a brief explosion of incredulity. "You filtered affluence data from the events depicted in my social tutorial." She made the observation as only an aside, then she settled quickly and listened attentively. Bear in mind she thinks very fast. We had packed water and beer even for our little train trip to allow us to keep our nanomachinery running at full speed.

"So, it begins. I am torn from the gentle life in academia or a bioengineering lab. The next day I must begin to reshape the world itself." She said this without irony. I would like to see her when she is joking.

"It could be quite fun," I responded.

"Too clear. It is too clear that your little internal storyteller is just jumping up and down with glee. As is my little inventor. But, I need a night's sleep, whether I need it or not." At last she showed fatigue. As adjusted as we were, we were not adapted. We could be overtaxed. It was something to remember.

We disembarked at Penn Station. The beads had distorted our time sense. In little time, we had covered a lot of ground while not covering much territory — or possibly the other way around. In any case, travel time was about an hour but seemed longer to perceptions.

Stepping off of the train I had a strange déjà vu. It was not an ordinary déjà vu which are naturally considered strange. This was more strange — less ordinary...oh well. As I stepped down, it was as if I stepped out of the train into the marvelous station of the early 1900's. Some wistful part of me was confused by the changes. I was missing the original nexus with its invigorating Jules Vernian tall-promises-of-a-prosperous-future-neo-classically-faceted-in-sweeping-arches-over-goings-and-comings that had started and ended so many of my little adventures in that century. The eerie sensation continued until I was out of the concourse. But when we reached the sidewalk, modern New York City air banished history and imposed a forceful attention to immediate needs.

We took a cab to the Westin Midtown. We left our bags with the bell captain. Then we took the subway a couple stops. Then I called a cab company to take our bags to the Kennedy Airport Hilton. We hailed another cab.

Melissa Accepts a Patron

"The Plaza please."

"That *The Plaza* you want? There's the Crown Plaza and a bunch of Plazas." We were advised by the pleasant cabbie/travel agent. Maybe it was our hats. We still had beads extended so we could carry on our discussions and our Investment Broker charade at the same time.

"Yes, *The Plaza* on Fifth Avenue, thanks."

"Isn't *The Plaza* a bit much? I have been using my credit card a lot recently." Melissa said.

"It is on me." She gave me a little thinker's frown. "Think of it as an investment." I said this with absolutely no distortion in my mental voice. I used my calm, confidence-you-can-trust-me-to-be-the-perfect-gentleman voice. "We'll work out the details later. Worst case you pay me back sometime. For now, all the movies say that people on the run shouldn't use their credit cards. In this case, I think they have it right. Suppose you burn yours, and I get you a line of credit to draw from and an ATM card. It is time for you to disappear."

"And you will burn yours?"

"I don't actually have one right now. I'll have to correct that tomorrow." This wasn't quite as she expected, but before she could inquire further, we were exiting the cab with the help of a nice white gloved assistant.

One thing that separates the experienced traveler from the rube, is conversation. In general, the less you speak to the helping hands of a fine hotel, the more important you are considered to be. And you are also considered to be a jerk.

We went to the registration counter. I started in with no salutation. "Leonard Schmedlock. I need two rooms for two nights."

In a quality, top-of-the-line hotel all systems are designed to cater to eccentricities. This person had never heard of Leonard Schmedlock, yet when the name was accessed in the reservation system a red flag meaning something like…"be totally super-nice to this customer and give him whatever he wants, no questions asked." You might say I was something of a celebrity to the computer.

"Yes, sir, we have something for you. Two adjacent rooms, two nights. Will there be luggage?" The astute and experienced assistant allowed for any kind of assignation without requiring a public declaration.

"They should arrive from the airport by cab shortly."

"We'll take care of them, sir. Here are your keys. Enjoy your stay. Shall we assist you to your rooms?"

"Thank you, no; we will eat first."

I called yet another cab company to fetch our bags from the Hilton and we went to get a late snack. I really just wanted to delay a bit while the room was readied and the bags arrived.

An hour later we entered our opulent rooms. The bags were there, the beds were turned down. The mints were in place.

"I'll leave my adjoining door open. Feel free to wake me for any reason at all. And enjoy this hotel. It has everything. It's paid for. Luxuriate while the opportunity is afforded you — a useful motto. Tomorrow is always less than predictable. Tonight, however, you will be treated like royalty, even though they have never seen you before."

"Thodkin," Melissa spoke out loud, "I'm glad you are here and thank you again for what you have done. I am a little amazed that I feel capable of sleep. Good night."

"Good night. Knock when you are ready to get started and feel free to call me Leonard."

That was it. We were in New York. We would deal with any investigation that might be brewing when we had better resources to do so. What could be done had been done to miscue any quick pursuit. It wouldn't fool anybody for long. But it would take a while to mobilize the bureaucrats. Then they would be after Melissa. Looking for Melissa at The Plaza would be an intuitive leap roughly the cross section of the Milky Way. There was time to shower and sleep. They have wonderful showers at The Plaza. I always treat a good shower as if it were my last. If I were on death row and they asked me to select the courses of my last meal, I'd try to trade them all for a shower at The Plaza.

A Sumptuous Breakfast to Launch the Changing of the World

Melissa joined me for breakfast. Neither of us slept late, and we were both still noticing small changes in physiology. As you move into your thirties, your muscles begin to marble. You are getting fat, but it doesn't show. I was no different. I do a lot of walking, but my eating habits tend to include too many foods high in carbohydrates and fat. For several thousand years I had slowly been tubbing up. In another 1200 years or so, I would have a layer of soft fat over my whole body. At about 36 or so for men, fat begins to collect subcutaneously. That's when it jiggles. But something was reversing my marbleizing. And it was doing it quickly. It was the dream of tubby America. And I noticed small changes in Melissa. Her general pallor was much improved. Live to be my age and you get pretty good at detecting who is healthy and who is not. I stopped predicting demise some time ago. It was no fun being right about that. And she had lost more soft curves…in a good way.

We left the hotel early and spent the day using money from one account or another to create a dozen new companies. Melissa figured out pretty quickly what I was doing and made suggestions for names and support businesses. We were about to go vertical. I was trying to cover the areas of expertise or service we were going to need. We would have to stock them. But I wanted to seed each organism as a newborn. This is partly because of who I am. Buying going concerns makes the measurement of accomplishment more difficult. In addition, even though Melissa and I brought unique perspectives to each company's mission, no existing management team would understand our single-mindedness. So we were going to start fresh. Between the two of us we could effectively filter for expertise and personality qualities. We were about to assemble the best people in the world into a very pointed change producing culture. Our science would seem like magic and to help us, we needed people who believed in making magic.

The truth? We needed magic to have any chance of success. More truth? Magic resides in the First Mind. We were looking for personalities already steeped in twinkle, already possessed by "maybe it is impossible...how is that a problem?" I once knew a garage inventor, that is an inventor who worked in his garage, not the other kind, who challenged IBM, giant of everything computational. He didn't know it was impossible to sling stones at the giant. In fact, his product went nowhere, and IBM didn't fall in bellowing thunder — that would be just too fantastic to believe. And besides that was the accomplishment of a few other people in a later time. But he built something, and it worked, and it rivaled the technologies of any monster of the time. And when he wasn't able to take over the world's most powerful computer manufacturer, single-handedly...he just kept on trying. Where is he now? Making magic somewhere you can be sure. We were going to seek people who could wait for recognition by the outside world. We must convince them of the bigger picture, the long-range view. We needed world-changers, relentless, questing and balanced. Failing that, we'd use what we could find.

You don't think myopic world-changing magicians are balanced, do you? I should say not. For one thing many of them are way too young to be balanced. They are under trained. If you looked at their personalities somehow manifest outside of the brain casing, they would seem like a colorful ball on the nose of a seal in a sideshow fun house where the walls had all been tilted to make the ball appear to be balanced on the side of the seal's nose, defying gravity. It would look like Fred Astaire dancing on the ceiling.

And balance is a dynamic process. Try to balance the pendulum of a grandfather clock. This is one of the great examples of the nature of chaos. The pendulum stays balanced only as long as not the tiniest unbalancing push is exerted on it. Then it falls prey to the weak force of gravity, which is really a huge, strong attractor. But counter gravity balancing is the personality of a hormone-driven young magician. Personality balancing requires motivation, observation, resources and action. Do you not value balance? All is unbalanced. If you aren't paying attention, you become *severely* unbalanced. If you don't have the tools, the wherewithal to adjust, you become more and more wobbly. And if you don't use your observation or tools to act…whoops! On your butt. You become another little tragedy of unfulfilled possibilities. Oh yes, there are many notable unbalanced achievers. Think of them as balancing a spinning plate on the end of a stick on a treadmill. They lean forward and run. The force of nature slides by under their feet and imposes running. Achievement happens in spite of the imbalance. We praise them. Just think if you could adjust imbalance instead of its adjusting you. Suppose you could use it and give it up? Suppose you could focus it and slough it off as needed? Sometimes it would help. Or not, believe what you will. We needed some magicians and a place to make magic wands.

So, Melissa and I sketched out a rough plan. Make that a rough non-plan. We needed something to use to lure unsuspecting magicians. If you don't know this already, money is of very little use when you want to lure magicians. Take a bunch of money. Wind the fishing line of a mediocre idea around the money and drop it into a pool stocked with magicians in a feeding frenzy and your tasty bait will not be noticed for all the other rich ideas in the pool…except by second-raters and self-touters. Magicians need the big idea…funding is nice, it lets them eat and if the idea is really worthy, there will usually be money in it somewhere. But, the true lure is a formula catalyzed by "it's never been done before." For now, we needed that quality. We had money. We definitely had the big idea. We needed people and places. Next stop… people. For that we had created Isus, llc and the ToolShops of Isus… nanoTs, geneTs, virtuallyRealTs, createdIntelligencesTs(ciTs) and the planning company conjurersTs.

The ToolShops of Isus

More on these later. It was now time to leave New York. It took me a moment or two — which is to say a fair amount of time — to convince Melissa to abandon all her worldly goods. We needed to drop out, disappear, go to ground, make a quiet exit stage right. We needed to blend in and become indiscernible from the backdrop of humanity. For me, that is a well established survival skill. Melissa was just a little surprised that she liked the new clothes and the Investment Banker shell. And she was so used to being frugal and worrying about paying back her student loans that she hated to waste money. But, I convinced her to let me call another cab and have our bags delivered to the United Airlines terminal at Kennedy Airport. We would never see these bags again. They would soon put them in lost and found. Someone with tools could match DNA, but otherwise there was nothing of us in the luggage, not even fingerprints. Most of the clothes were brand new and unsoiled. Anybody following us would track those bags, scour the airport manifests after some expensive bribes and find no Melissa and no Leonard Schmedlock. Melissa was a known loner and geek. Leonard is known to take long vacations from the law firm bearing his name: Leonard Schmedlock, Law Booth. Indeed, nobody currently working for him had ever seen him face to face. Those who had teleconferences and video conferences could be relied upon to yield up differing descriptions. He wasn't known to travel with a companion and had in true fact never met Melissa before one day ago. But, just in case, we would be taking mass transit up state. We would follow a winding trail and in a month, not sooner, we would get to Boston. One thing that age might teach you is patience. We needed some, and I have it. I've been waiting for humanity to come to this place in history for at least 9,000 of my 10,000 years. Now was no time to be hasty. And besides, we had things we could accomplish on the road. But first, we needed a Salvation Army.

New Shells

Walking into a Salvation Army in good suits, well coiffed and with rosy cheeks is bound to bring sympathy. The Salvation troops have seen all kinds of failure. We were just another case of big ideas and gluttony gone bust. They allowed us to trade our clothes for more modest and versatile Levi's and sturdy winter coats. We told them of our desire to quit the rat race, and how we just couldn't give up our belief in the American Dream, and how we had bought a Falafel Wheelie Stand with money from our Rolexes and staked out a corner in Midtown, and how we thought we needed clothes that were more practical and how it would be good to get closure by giving up the last vestiges of our former life. We told them we didn't need the shelter, and thanking them for their concern, departed heading towards the flophouse part of town. Out of sight, we took the subway to Penn Station and caught a train heading in the general direction of Buffalo.

Far Away a Plot Develops

You may have been wondering if everything was just going to keep going our way. If you ignore the untimely demise of the AOU... again, and the interruption by the President of Princeton forcing us to flee the ivy halls and hide out at The Plaza, you could be thinking that everything was just too easy for us. And I would find it possible to agree with you, although the time seemed somewhat busy with moments of uncertainty. But in fact, I had not predicted the President's actions to any great level of precision. Yes, she had taken some time to deal with the bureaucracy of someone dying in her realm... she was more like a queen. But she had determined that this was one of those rare opportunities to forward personal goals. She was tired of the confines of her job. She wanted real power and that required real money. The plain fact is that not having money keeps a lot of evil-doers from evil doing. She was secretly scheming to become corrupted absolutely. In order for that to manifest, first came absolute power. Phase 1 — power corrupts — had been accomplished. For some reason, her natural skepticism, which she had built around her like a massive stone fortress, had been breached by just a moment with guileless Melissa. She had believed her eyes. And having a prefab green monkey try to shinny up her leg had solidified her resolve. She was going to find Melissa and somehow milk her of the treasure she was hiding. She was going to lock Melissa up in a milking stall, attach an udder pump of lawyers and middle techs and suck the treasure from Melissa, night and day. How this was to be accomplished, the Queen didn't precisely know, but there was a sporting expression cycling through her mind: whatever it takes. In sports, this typically refers to personal sacrifice; that was not her thinking.

The Queen of Princeton Enlists Minions

It is an unfortunate by-product of social pressure that behaviors are developed and accepted that are fully counter to intended purpose. Let me state that another way. The stress of society builds up like geological plates. When geological plates push together at a fault line, the pressure can build until it is abruptly released in an earthquake. Society is very similar. Because society moves so slowly and with such tremendous inertia, it is without analytical awareness that pressures and counter-pressures develop. The most common form of this is between intention and actualization. The most devastating is of the nature of "do as I say, not as I do."

All children are taught about rightness and wrongness. They are taught that rightness is, well, right. And they believe it. Almost immediately upon learning how to judge between the two polarities, they are taught that these are ideas that just don't work in the real world. In the real world one is bounding toward failure the moment one tries to actualize rightness. One will never be able to live…rightly. This is known by everyone, but what is not perceived as clearly is that the idea pervades society and all the machinery of society. A mundane example, yet germane to this discussion, is that youth are taught that it is important to go to college in order learn to become well-rounded social animals and about history and their special interests…and besides, they need a degree to get a good job. But getting into college means getting good grades. This is complicated for many young people because an artificial specification has been developed that defines the time allowed for the achievement of the objective. You must learn in lockstep. If you can't, you must learn how to appear to learn. There is no such thing as learning at appropriate stages of development. Maybe when you get to college it will all be different? Of course it isn't. The forces building between the intention of education and the mechanism enforced by a reactive social establishment will ultimately result in a chaotic adjustment….

But, for the moment, it was a useful thing for the Queen. She just needed to find one of the formula kids. One who was effective, but couldn't think for his/her self. One major kiss-ass. One major do-what-it-takes-to-get-a-grade kiss-ass. One kiss-ass with the right major. If you are one of these, and have justified it in some way, I hope you aren't too offended. I am not saying it is your fault that the pressure exists, just that you may be better served to seek alternatives to giving up your honor and integrity. And let's not mention abandoning any really useful education because of the nameless fear imposed by a social force. And if you find yourself in the grip of the Queen, I'd say run for it. She is not a good person, and she will not treat you well.

The Queen found a pseudogeek, not quite living up to expectations, but able to employ some elementary wiles. She told him that she had an extra-credit assignment for him, and would he please help her find her lost niece. She told him that Melissa was her niece and had burned out from the nameless pressure and disappeared. Could he find her?

Chronology of the Search for Melissa

I could go into arduous detail about how the following events were executed. I have gathered detailed accounts. But this has all been done so often that I find it a little tedious. Let me list the process in outline so we can get on with less predictable history more germane to the topics under consideration.

CI_132 L0ARII:

The President and the Pseudogeek form an unbalanced team and jointly make the following reality:

Failure	Pseudogeek tries to find a program to access banking databases
Failure	Pseudogeek asks around for Trojan Horse software to infiltrate banking databases — contacted persons are too busy studying to take the request seriously.
Success	Queen finds Melissa's credit card number in her University files

Failure	Pseudogeek goes to Melissa's apartment to search for clues
Success/Failure	Queen impersonates Melissa using college records, discovers Melissa has not used credit card since she disappeared.
Failure	Pseudogeek fails to discover Queen is only child, has no nieces
Failure	Pseudogeek leaves finger prints in Melissa's apartment. Missing persons investigation instigated by Queen will add Pseudogeek to suspects list.
Success	Queen denies asking Pseudogeek to break and enter. Shows fallacy in geek story fabrication...explains she is only child. Pseudogeek expelled on morality grounds.
Failure/Success	Pseudogeek goes to junior college in Los Angeles, hangs with friends at mall, manages to get pants-waist to hang just below glutes and above penis at a rakish angle, buys too-phat boxer shorts. Parents refuse to pay for books and tuition. Pseudogeek begins self-discovery adventure and turns into caring social-worker-Luddite. Time span = 11 years. World is well on way to new-dynamic. Geek sees IsoTown prototype, applies for citizenship of newly formed "New Town of Total Sociality." Ultimately rejected because of history of breaking and entering. Deals with disappointment of real world. Thanks nature for new lesson in the importance of understanding consequences of action. Removes telephone and forces co-workers to send schedule by physical mail.
Failure	Queen's preliminary efforts turn up nothing regarding Melissa's whereabouts.

Bear in mind that the above failures were due to a momentary underestimation by the Queen. She had underestimated Melissa. People in positions of power grow insular. They lose knowledge of the potential of the average person. And, of course, Melissa had me. That is the conjecture that occurred to the President. She decided Melissa was not worldly enough to disappear effectively: someone helped. I was suspect, just because I had been on and then off the scene. This and one other thought improved the Queen's chances of success. For the first time the full potential of this opportunity sank in. Imagine the wealth and recognition to be had from the proper use of, well, whatever it was! This was no time to spare horses. She called an investigator. They do exist, and some are unbelievably good. Believe it. I was pretty sure, as we sat on the train to Buffalo watching more and more snow accumulate along the tracks, that somewhere the Princeton Queen was drawing upon the power of fealty and casting her evil eye in search of Midessa and her Au-laying brain. My goodness, that is a hodgepodge of analogy. But so true. When I was younger, I used to say "some things never change." Now, I say "some things rarely change." And, I am afraid at this particular moment in history, the Queen was just a storybook cliché. Maybe her methods would turn out to be interesting, but her motivation was simply camp. She had started well… American Dreamer, Overachiever, Leader. It was sad her training was limited in the "purpose" context. The good thing about her technique of evildoing was that she only had contemporary fantasy to draw from. And contemporary fantasy has innumerable restrictions. The P. Queen had self-manifested only a paper-tiger divine right. So, she was not really able to live up to her capabilities, at least early on in her quest. We'd give her a little time to grow into the role. She had motivation, and she wasn't used to losing; that was a start. For the present, I was thinking no offensive strategy was required. We had a pretty good lead, the clock was on our side. I reaffirmed the fade away for near-term tactic.

The Cleo Award

Queens from whatever walk of life always remind me of Ptolemy XIII and Ptolemy XIV. I was once a waiter for Ptolemy XIII. I was a very good waiter. Good waiters demonstrate respect and animal wisdom, not unlike bartenders. Ptolemy XIII told me in one of our conversations that he never wanted to marry his sister. He feared it would end badly. Of course it did; not long after our discussion, he was killed. Cleopatra then married her younger brother Ptolemy XIV. But she didn't really like younger men. It seems that once you give yourself, wrapped in nothing but a rug to a Caesar, you start to develop cravings for Italian…domination. But, anyway, Ptolemy said that certain people seek power because it is the only game that makes living fun.

Of course, Cleo ended badly. She was designated a sorceress by the Roman Senate…all men and a predictor of the future, where powerful men would demonstrate piety by abusing and persecuting women. An ambitious, privileged woman, she was no judge of alpha males and picked a series of losers, primarily because she didn't understand a basic tenet of power. It shifts. Queens resist the shift. They are so interested in staying Queenish, however they define it, that they attempt a very difficult goal. In order for one thing to remain the same, it becomes necessary for a large set of other things to be changeable. Using other words, Cleopatra adapted as well as she was able. Remember her family had been political for 200 years. She was bred to politicking, but was mislead by the success of all those Ptolemy numbers that came before her. She never understood that the rest of the world would have to be controlled in order for her to stay Queen Philopator and Pharaoh. And there was no other job she knew how to do…she chose self-expiration. In the end, she was powerful in some contexts but could not change well enough to provide for her own survival. So, if you are rooting for the Queen of Princeton, listen to Ptolemy… Ptolemies. They knew it was a bad idea to get in bed with a Queen.

For Want of a Meme, the Egypt Was Lost

When Cleopatra VII ascended the Egyptian throne, she was only 17. She reigned as Queen Philopator and Pharaoh between 51 and 30 BC, and died at the age of 39. My analysis indicates a critical personality flaw infected the Ptolemies' dogma in less than 200 years. So severe was this flaw, this meme space, that within 17 years a 200-year-old tradition became extinct, eliminating not only all power supporting the Queen but the life of the Queen herself. Surely, her education — specifically her understanding of personality — was flawed…and she was not alone. When you follow any

"Strong, unchangable ideas—those are Turtle Ideas.

Beware Turtle Ideas"

—*Thodkin Marblemeister*

rigidity you become subject to its attendant fragility. When dogma is exposed, dogmatists and their followers fall. I don't think Cleo had very good coaching in this area. And think about this for a second. She married her eldest brother; he was very young and lost a war. She married her younger brother; he was killed in a duel. She was 22 when she hooked up with Caesar. These were all kids. I am not one to deify history and project its repeatability unquestioned, but these kids could have used some instruction on how to keep a strong dogma strong. Egypt never again reigned. Seriously, it was the kids and their poor education regarding the law of requisite variety that caused the fall.

Specifically, Ptolemy XIII said to me…"Waiter, please get my sister/wife another goblet of wine before she kills me." I fetched it barely in the nick of time. I always thought you shouldn't marry your sister, the why of it became clear, although Ptolemy XIV must not have been paying attention — brother XIII had shown the sister to embody ill-omen. I saw the handwriting on the wall, quit my job and found employ in a small cafe near the Coliseum. It wasn't the best Isicia Omentata* joint, but there were franchise opportunities, and people weren't in such a hurry that they couldn't listen to a good story.

Now, it is true that a Queen at the peak of power can produce temporal disruptions. But time extracts its own plodding tax. If a strategic plan of sufficient span is implemented, the resources of a Queen to cope with the elements of a large system are bound to become exhausted. I wouldn't want to be in the Queen's slippers when her minions discover she sleeps on a bed of empty ideas. Remember Cleopatra, or challenge infinite complexity. You choose.

The Law of Requisite Variety, a short course:

- Concept credited to Systems Theorist W. Ross Ashby.
- Concerns the amount of flexibility or choice an organism or member of a system requires to successfully adapt.
- Adaptation most commonly allows survival…no adaptation brings extinction.
- Flexibility must be commensurate/proportional to the variety present in the rest of the system.

•••••••

* A Roman burger made from minced beef soaked in wine, some pepper, Liquamen, stone-pine kernels and green peppercorns and Caroenum — served on a roll.

Boogie to Buffalo

It turned out that the AOU was also a boogie-woogie freak. I had quite a collection of tunes wired into my head. You are aware of the "Walkman/iPod"? I am forced to predict rapid obsolescence. Sell your stock. I could now carry on a conversation and have background music at the same time without disturbing anybody. So I was doing some planning with Melissa to a nice walking bass that just fit the train's rhythm.

After a period of contemplative silence, Melissa chose to summarize her thinking. "We are headed to Buffalo. In the middle of a not-rare snowstorm. That would be to confuse our trail? The snow will cover the tracks of our horses, making it difficult for the posse and confounding the Indian guide? I think it should work. Were I in pursuit, I'd be thinking someone on the run would tend towards increasing options as opposed to being trapped in a corner of the country." Melissa was applying her intellect to the idea of being "on the lam."

"Follow rigidity, become subject to its attendant fragility, it is just that simple."
—*Thodkin Marblemeister*

I sought to clarify. "I have learned that consultants prefer the warmer, more fun climates. We can expect any professional the President enlists will begin the search in something of a quadrilateral with San Francisco, Los Angeles and Miami as conjoining vertices with Princeton. And you have family in San Francisco?" She quickly remembered I had access to the AOU's bio of her.

"Yes, I see your point. She will certainly need expert help. And we know that finding your family would require archaeological skills… no offense and forgive me for any sadness my being blunt causes. I'm a little better at discerning when I should be more sensitive, but my mouth is still running ahead of discernment."

"Please, I prefer your unexpurgated style. By now, I have a certain skill at managing inquiries about my heritage. Your family in San Francisco will be the first search area. But an equal imperative will be the consultants' personal desire to just get out of the rain at the client's expense. As for Buffalo, I wanted to go to a city with a good college that you have never interacted with. I researched your publications and collegial relationships and found none were currently in the Buffalo area. We can expect anybody you have ever known and any place you have studied to be a search area. The stay-at-home quality of your life will actually make it a straightforward task for adversaries. I won't care about that in a month, but for now, we need a place you can blend in and maybe get some work done."

"When doing my work, I'm going to be somewhat conspicuous if we don't find a 'front' for me." Melissa had thought this through on her own initiative. I needed to anticipate her better. I was not yet used to running on three timelines. You might think that a day is different if you have lived 365,000 of them. Actually, a day is just as long or short as my ability to get into *flow*. That is the same for you. But, when I review distant history, time seems to speed up. This appears to be different from those with shorter lives, who have a sense of time more like an accordion behind them. The high points, the low points, the significant events are connected with less descript smooth fabric. My most recent 100-200 years are like that, but 5,000 years ago speeds into my mind almost like a re-enactment done with animated flip-cards. All the tweening is eliminated and only the "news" or required context for the "news" is noticeable.

"I was thinking Buffalo might sequence this way…

"We find adjacent motels. We rent you an apartment. I'll take the 'story-teller's' way. We will rent a work space near to UB capable of covert entry. From this space we could set some transitioning goals and abandon Buffalo when we are ready."

"Most succinct, if also without any real detail. One thing I am going to need soon is a task. I don't like being widely focused. Have you given thought to how my knowledge can be employed?"

"Yes, a little. We need to bring about an acceptance of a radical new technology. We need to do something that will be acceptable to the population as a whole…something grass roots can grow in."

"You are saying, we shouldn't produce a Green Hair Comber Mark II?" Well, I had just heard Melissa make a joke. It was accompanied with a smile. She had a nice smile, which was somewhat surprising. How does one develop a nice smile without smiling? And it is easy to tell when someone is smiling, but not really smiling; people want you to know they aren't smiling so they betray a pure smile with small flaws. This was a genuine smile…I again raised expectations for her.

"No, but now that you mention it, I can see you are right. How about something outrageous, but depletable? Something that must be replenished, that is similar to something already in existence. Something the government would not immediately ban. Something familiar, friendly, but not intelligent."

"I'm a scientist, not an engineer, much less an entrepreneur. I don't immediately think of products." She did want to do something, though. Something she could point to and say "See, you bigoted, sexist manipulators? This is what I can do." I was going to have to protect her from those sexist bigots. They still existed and just loved to burn sorceresses at the stake…some things change quickly, others slowly. But, no short course in bigot survival was going to work for this. She needed to figure out some of these tricks the hard way. My experience would work as a buffer only for a while.

"We can get help, but let's take some time to address the issue. Maybe you are more of both scientist and engineer than you know. Perhaps you can use a little of your Day Dreamer thought space to invent Melissa Engineer. Entrepreneur is simply a matter of recognizing a need or desire and implementing an effective strategy to bring it into existence…and make a few bucks. By example, let me ask the question this way…what was the first fun thing you remember doing?"

"I don't remember." She quickly responded.

"If you could remember, what would it be?" I prompted.

"I don't know." She replied.

"If you make believe, just for a second, that you do know, what would be the first fun thing you remember, or nearly the first you remember now?" I continued to prompt.

"Coloring. I remember getting crayons and being amazed by them." She found a way to the memory after all. A note: typically "I don't know" from youth really means…"I'd rather not say, because I don't want to expose myself to a sneak attack from a wily old fogy." She had eventually found that she felt enough trust to answer the question. She mused on crayons, for just a split second before moving on.

I made a surprising observation.

"I think you have just chosen something that meets the criteria. Everybody has loved crayons at some time. With the exception of parents trying to keep children from drawing on walls, or eating the orange colors, you would not find a single scrooge for crayons. What can we do to make crayons better? So much better that they seem like magic?"

"You want me to invent a new crayon? I could do that. I am reminded of something Marshall McLuhan said about the use of television. He said TV was equivalent to movable type only being used to make comic books…or some such thing. You want me to use an unbounded ability to change the world to make, instead, children's coloring sticks." She mused upon this, and had not said it with sarcasm.

I felt an expansion was necessary, "McLuhan observed that technologies reinvented people. He said they were used in a pressure for homogeneity, an evolutionary pressure to bring about a single point of view. He argued we are tending toward a common representation of experience…if I can interpret his thoughts.

"I think, instead of enhancing homogeneity, what you have is the capability to stimulate the exact opposite. It is the ability to produce cohabiting variation. Reinvent people yes, homogenize, no."

"Yet, you think we should insinuate, versus overtake. Reinvent people without RIFT. I admit that less than a week ago, I would have laughed at you as a coward. Understanding stemming from recent experience — and I notice something of your social tutorial — forces a somewhat different response now. I agree completely. I have been fighting RIFT for days. I wouldn't wish that battle on my worst enemy…if I had any enemies. You seek a gentle revolution. Why not?" She was the most brilliant person in the world. You will notice that she adjusts very quickly when provided adequate information.

I offered one more observation about change. "Most commonly, change is adopted in a pyramidal progression, starting with small numbers at the top. Many contemporary-cause celebs have traveled from the top of the pyramid down. There is a simple explanation for this. The individuals most susceptible to an unorthodox idea are the people who already understand something of the context. They tend to be few. The more of the context they understand, the easier it is to adapt to a well conceived case for change.

"The second factor is that the people best able to communicate the case for change are the people just above you on the pyramid. Not only do they understand the context, but they are closely related to the next layer. It is difficult for a lay person to understand esoteric mathematics, etc."

"I see that in combination with your much touted nail-that-stands-out thesis — it would behoove us to have the lower levels of the pyramid firmly reinvented before the dogmatists understand the far-reaching implications of the inoculations." She summed up, "If I were in a position of power at the top of any common dogmatic, I might be scared shitless."

"Dogmatists are always scared of something. Succumbing to the fear is a necessary supporting pillar of dogma. Recently, a children's book was burned in effigy. This book was the single motivation for tens of millions of young children to re-include reading via internal impetus in daily experience. It was burned because the story environment was a fantasy depicting a benign magic as a plot tool. Many valuable and generally supported nonsecular subtexts existed throughout the story. Yet, a collective of endungeoned minds — a religious order — made public declarations describing the stories as evil. The leader admitted to never having read the book. What was the fear? Were they afraid children would suddenly be casting spells and doing evil? No, they were afraid their children would read, and during reading, think, thereby breaching the dungeon the dogmatists were attempting to patrol."

“This is likely to happen to our work,” Melissa remarked.

“I’d like to try a subtle approach, first. We have other options. We will need more than one ‘front’ for you. Could be fun. I’ve become a little intolerant of this long-lived ability of the hindbrain to collapse in upon reason.”

“I can understand how you might feel intolerant. Understanding why you are not insanely intolerant is more difficult.”

Stopover in Honeymoon Heaven

At Niagara Falls, Melissa and I split up.

The sinew — the connective tissue attaching the present to the future — is organic. It is grown from bonds formed during changes and fertilized by the detritus from elements disassembled, dissolved into potent chaotic noise. A description of one such maturation is being described. There are millions in progress now. Some will shrivel before they help to articulate the levers of the future. Some will go to Boston, others will find themselves in Buffalo.

The train ride takes about 7.5 hours from Penn Station to Buffalo. But we detrained in Niagara Falls and stayed the night. I'd been there before, but it was prior to the civil-engineering job to prevent the falls from falling. In detailed fact, it was before the town had running water other than the falls itself. It was a fun place to do a little nature watching. It was beautiful then and I was interested in making a comparison. Besides, it was a jumping-off point for Canada. If someone managed to track us this far they would then have Canada to consider. They were certainly going to think we were running far away. They would be thinking far away.

Much of the train conversation took place in quick time, beads warmed from information transfer. Eventually, Melissa went silent and began to ponder the possibilities of coloring sticks. I began to outline the future. It is best to design your future. No matter how it turns out, you will be better prepared. And it might be best to think of a design as a variable — things always change. Too often a plan is considered an invariable specification. The moment it inhibits a range of behaviors is the moment it eliminates possible outcomes. So many wonderful ideas and inventions have sprung from unrelated tasks and goals that it amazes me that plans, designs and schedules have become nearly synonymous for concrete. Just think of formulas as recipes for reproduction and never use them for strategic thinking or planning. If you find yourself wanting to argue that deadlines are needed for coordination of complex endeavors, believe what you like. True immalleable deadlines are extremely rare and almost always are a result of inflexible, limited design processes. Think more first, scramble less second. But no amount of forethought will eliminate the need to scramble. Both abilities are useful, planning and improvising. And remember, soon the human life expectancy will extend so far beyond its current range as to be considered infinite. What is a deadline then?

While I'm Planning Melissa is Dreaming

Melissa reported her waking dream to me. "I was wearing a beautiful dress with no underwear and had a somewhat more voluptuous figure. All around me people were moving, making little twitching motions with their hands and never making eye contact with me. The motion of their bodies was forcing me backwards toward whatever their destination was. As I moved, I struggled, but eventually I was pushed into a huge greenhouse cavern. The twitchers quickly faded away to little work stations where food was being processed. Finally released, I took an elevator, a vertical belt with alternating foot platforms and handholds, up several levels. I kept riding until I was out of the greenhouse. When I stepped off, I was in a sterile breeding nursery. There were women's torsos hooked up to pristine machinery. Many of the torsos were pregnant, others had babies at each breast. The nurses tried to pull me away from the lift belt, but I got free and jumped on again. I rode up further and soon came to several layers that were exact duplicates of a rich New York flat. I stepped off the belt. The lift belt was hidden from view of a fancy elevator. From behind an ornate Japanese barrier I watched women wearing my exact dress and built much like me laughing and cajoling a variety of men into the apartments. These women were diversity lures, keeping the hive from becoming inbred. One of the women saw me and furiously waved me away. I noticed my head in a hallway mirror. It was huge and bumpy. I had no hair and my cranium was the size of two medicine balls. Only in a dream could a neck my size have supported such a head. But my pheromones were being released in proportion to my head size. Every man who entered the hallway whipped around in my direction and took an involuntary lurch toward me.

"The dream demonstrates an evolving self-reference. When I was a little girl I devoured science fiction. I salivated over the worlds and ideas of clever conjectures and commentary. I could barely converse with people who were content with their present. I remember reading *Hellstrom's Hive* when I was quite young, a Frank Herbert social future exploration novel. It described an alternative socioeconomic system modeled after bees. There were two horrors in the book that wouldn't leave my mind. One was that I never wanted to be a drone — programmed meat. And I never wanted to be a breeding stump. But now, I have crossed over into a new species altogether. I've been turned into a seducing bighead. It was a disturbing exploration. I'm convinced that Frank Herbert was wrong. A hive model will not succeed. More specifically, it won't because it can't evolve. If diversity spontaneously jumped too far, it would be subdued by a conservative establishment whose purpose implies eliminating the incumbent dangers of change in favor of order. Any potential for radical change would be burned away. If we need drones, then drones must be made from something other than human organisms, and we cannot de-rate humans from individuals to breeding ovens."

Her dream is communicated to me fully accessorized. We now have the ability to communicate strong nuance with a standard codification. I can't say I enjoyed her experience at all, but there was an important message here. We were going to have to do a very rigorous system design, in fact a set of design alternatives. We were going to run the risk of tripping over primal objections. What is wrong with being a drone? Drones are drones, functionaries without awareness of alternatives. What is wrong is de-evolving toward drone? An argument could be made that we were already well down that path. But, I was thinking we'd better find a way — several ways — to ensure humans turn away from it.

Not that Melissa was in any danger of being mistaken for a drone. Nobody would have guessed she was a bighead. Her body changes were more lurish and she seemed to have regained a coordinated physical smoothness that had only been hinted before. Nothing about her would qualify her for the Drone Corps. She was safe from her dreams being fulfilled. At least from any Hellstrom misqualification. As for being safe, I had begun to think of her as Earth's prize human. I was pondering her safety with more imagination. But it is not really possible to protect someone. They must share the task and even then it is only possible to improve the probabilities of survival. She was going to have to get good at designing her own safe passage for the coming years.

Rent? Or Own?

I phoned the Leonard Schmedlock offices and had them arrange for a living space near the UB campus. After legal and accounting analysis by the wonderful minds working for Leonard, it was determined that a quick-escrow purchase of a house was the best approach. So, instead of a minimalist apartment, Melissa spent a week in the Niagara Honeymoon Fall Forever Motel complete with tragic jilted-at-the-altar account, augmented with Salvation Army wedding dress prop, which she chose not to talk about. Secluded, visited occasionally by a commiserating room service and uninterrupted by maid service, she began to think in detail about coloring sticks. A week later, she quietly left the motel to "go back home" amid a series of well wishes from the motel staff, all translated as "You'll be fine, dear. He was a louse anyway; you are better off without him." As a nation, Americans know when to offer condolences, even if the dialog is a bit familiar. Oddly, Melissa said she felt better because of the genuine concern she felt from what could have been a jaded staff.

Later that night, after a series of misdirections learned from Agatha Christie and Arthur Conan Doyle, she knocked on the door of a fully furnished mini-mansion that was to be her home for the coming month. I had arrived a bit earlier and entered through an inconspicuous kitchen delivery entrance. I let her in. She laughed at the solution we had employed to secure a hideout for her. She was to live in seeming wealth. Let the Queen and her hounds guess that. If the trail didn't lead them directly to us, it would be a stretch to knock on the door of every recently sold house in Buffalo.

I went to Buffalo and was lucky to get a job as a temp. I was temporarily a member of the UB janitorial staff. This is a job I understand and I had excellent references from, well, the Law Booth of Leonard Schmedlock. My job was to find suitable candidates to assist us in some small engineering tasks. We needed to be connected. We needed access to global information and resources. We needed portable, secure high bandwidth interfaces to as many communications channels as we could manage. We were oddly quite blind. We had eyes to see, but nothing to look at.

Fortune Smiles on the Prepared

It is true that Art Seidleman was lucky because he was prepared. And he was lucky because he placed himself in the way of opportunity. And he was lucky because he spent much of his time in environments that might be termed luck pools, or opportunity pools. Universities are luck pools. And if you are prepared — say if you have lived with a variety of people and paid attention — you will be able to siphon some of the luck out of a University Luck Pool. I once heard a high tech manager explain his secret for picking talented individuals from the stack of résumés he would receive for a job opening. He said he always asked what personal projects they had created. This was particularly useful if picking the best recent graduates who tend to have carbon copied curriculum vitae. It was a question that would weed out those who thought a profession was a good career move from those who were compelled by the endeavors of the profession. We needed a super-geek who was motivated by more than "the right career moves."

UB specialized in a set of research areas. The most promising at the time was Information Fusion research. This is something that Melissa could boost to our advantage. I began by hunting down the key students with electrical engineering or systems backgrounds. If forced to, I could fill in as an electrical engineer — a hobby of mine — but I was going to be taxed by other responsibilities. We needed to co-opt some native talent — talent with access to facilities. We weren't going to hang around for long or we could have made use of some other UB areas of research and some quality equipment. But, we were on a hit-and-run mission to get connected. We needed to fuse some information. We actually now had minds that could cope with simultaneous variform detail. If we had access to better, higher-quality information, we now had capability to use it. In the spy trade it is called intelligence and is used to improve design of covert operations. We were just facilitating the process of implementing a plan. A plan not fully fleshed out, but sketched out.

Melissa settled in and we met every day or so. I volunteered for night cleaning jobs, which is good, because that is what they give temp labor anyway. This allowed me to see who was working overtime in the labs. I narrowed down the field to two people and finally tested the first.

Pizza Is Not Just a Place in Italy

"Hey, man, mind if I eat lunch in here?" I directed my conversation at a skinny black kid surrounded by a variety of oscilloscopes, lab equipment and prototyping tools. Little curls of rosin smoke drifted up from his lab bench. He was soldering something together. I had already learned that he was building a very-high-speed optical interface for one of the multiprocessor arrays the University owned. But, for now I was just the janitor, looking for a secluded place to eat my night lunch. I had a very large pizza and six-pack of Coke with three cans missing.

"Sure." He didn't look up. He had seen me mop the lab floor every night he was there. I was no surprise to him.

"Cool." I sat at a corner situated close enough to see him, but not impose on him. The pizza quickly overran the soldering smells. I popped a can and started in on the pizza. "Hey, I got plenty. Want a piece?"

This got his attention...a brain needs to eat. "Yeah, I would, the smell is nukin' me." I held the box out to him. He wiped his hands on his pants and picked up a piece. Never look a gift pizza in the mouth. I could have had a pizza made from pigeon droppings and he would have taken a piece.

"Have a Coke."

"Cool...thanks." He was young, but not so young as to think that a janitor was going to be stimulating company. So, he gobbled a really big bite and went back to work. I was going to have to get him talking.

"Looks like you are doing some high speed stuff."

"Yeah, really high." I hadn't cracked his pride open yet.

"So, how high…gigahertz?" This finally got his attention.

"Sort of. You know about this stuff?" This was about as subtle as he could be. I must be quoting something I'd read or heard while mopping.

"I know a little. I read a lot. What do you mean, sort of?"

"I'm building a word-wide optical interface. Each fiber is a little over 1 gig. The word width is 128 bits. So it is really 16 gigabytes." He figured that would shut me up.

"You using gallium arsenide chips?" Now I entered his secret territory, and his territory of pride. He had to expound.

"Actually, this uses a new Bucky tube technology — very new. You know what that is? Carbon nanotubes? I have created my own 11-nm device - GaAs MOSFETs are lounging around 11nm, won't go much too much better than that. I'm starting at 11nm. I had to fab them myself with this cool self-assembly process. It has never been tested except on the simulator. If this slow configuration works, my next device let me create a network that processes at or beyond the Petabits per second of a multi-core fiber. Basically compute at the transmission rate of the network. Big data is going to get way bigger. Massive data." Now his pride showed. This was something he knew nobody expected from him. But he had tired of moving at expected pace and had been spending every free moment on his skunk-works project. This was the super-geek I was looking for, at least in motivation. Something to prove, and something to prove it with.

"How many processors in the array?"

He lifted his head from the soldering for the first time since taking the pizza. "You really a janitor?"

"Janitorial Engineer…I am right now."

"You really must read a lot. Maybe you should go to college."

"I'm just curious, how many processors in the array?"

"One hundred twenty-eight. The node memory is too slow for testing my top end — I could do more if I had a better lab."

"So, that lets you prove the architecture. And you can derive some metrics on the buss sharing of a real world array with real tasking. What then? A RISC node?" Now, I had his full attention. We weren't peers, but I was allowed some unbegrudged explanation.

"Yeah, but I can't afford to build a processor, so I can goose the array, but I'll be processor and memory bound with what I have to work with. Some day I will make my own node. Maybe I'll get a job with IBM. This should interest them. Nobody has done this before. Certainly, no sophomore…and without shit: no grant, or anything."

"I haven't researched the field, but it sounds very impressive to me. I think I read IBM is working on nanologic and Buckytube transistors."

"Yeah, we don't have anything that can do that here - had to make mine myself. Takes a big grant, or a military project to research that stuff in a college. I just hope I don't get there too late." There are times when breakthroughs are imminent and it is a long time between that breakthrough and the next one. It takes a while to make breakthroughs pay out. If you are in the payout period, it can seem that recognition is unattainable. Mostly, beginners — kids — think that. The kid was feeling pressure to be recognized as an innovator. But he was self-propelled. His little beany was a twirlin' from the thermals his brain was giving off.

"I'm sure you are going to get noticed. Well, I gotta go…later - you take these." And I left, leaving half a pizza and the last Coke. Least I could do.

For those of you who are not current on state-of-the-art computing technology, we now dwarf the kid's numbers. But at the time, his idea was leading edge. A processor array is just a bunch of computers able to work on the same complex problem at the same time. Being processor-bound means you were kind of starving for computing power. In Melissa's and my own heads, we had processing power that could out-compute the whole University 10,000-fold. But, we were I/O bound. We couldn't get information in or out, except to each other, or through our fingers, eyes or ears. The kid was working on exactly the problem we needed solved. We needed to link to the outside world, even if it was very slow. We could always speed it up later.

I spent the rest of that night pulling the kid's design from the simulators database. The next day I communicated it to Melissa. She agreed it was very good and gave me a green light for the technical side of the kid's credentials.

I made another phone call to Leonard's Law Booth and had them start hunting a facility for *nanoTs*. We would use this as a foil for subduing our hardware obstacles. I was going to start populating it sooner than I expected. You find talent everywhere. When you do, jump your pins in front of it if you can, the bowling ball of opportunity is rolling at you.

The Truth about Executing Logistics

Rather than explain every detail of the several months following the conversation with super-geek — he was the first of many — and to avoid the risk of drifting this little story, now chronicle, off topic, I'll step over the month of rapid staccato successes. We spent a month in Buffalo implementing mechanisms for interfacing with the rest of the computing world. SG1 was able to do in a month what his whole department couldn't have accomplished in a year. What he created was a little crude and would be upgraded but was critical to our next steps. In the interest of not boring you with the mundane, let me outline our quick steps. First, Alphonse, a.k.a. SG1, dropped out of University. Rumor is that he was hired by a top-secret lab, somewhere in…well, somewhere.

Melissa Speaks to Woman in Santa Fe

But first a precautionary note, lest you think this was all too easy.

The Doogie Howser Syndrome

- What you see is what you gotta get.
- What you see is what you want.
- The best things in life are free.
- Somebody else has the best things in life.
- Hard work and learning are just other words for lucky.

In the society of the times it was considered obvious that young people were losing the ability to work for their desires. This, of course was nonsense. The truth is that some were, and many weren't. Another truth was that the syndrome wasn't confined to youth. In fact, whole nations were expecting immediate gratification. What you see is what you get had taken on a new context. Everybody had television. Everybody had access to Hollywood fantasies. And, for various reasons, these fantasies distorted reality. Who wanted to watch reality? Even "reality programming" was about watching other people seek unreality. So, the world as a whole, began to succumb to the Doogie Howser Syndrome. For those of you who don't immediately understand the reference, Doogie Howser was a fictional character depicted on a weekly television pseudomorality show. Doogie was a boy genius who became a doctor. Each episode concerned his troubles coping with being a child doctor. That is, more precisely, he was an internist, but he was a child. The problem with this show was that Doogie achieved everything, but we never saw him work for anything. His ascendance is managed in 15 seconds of back-story narrated during the opening titles. But, we know — that is we who have things or knowledge know — that everything good takes effort. Don't listen to anybody who says the best things in life are free. The best things in life, clearly, are paid for with effort. Even geniuses go to school. Show me one genius who was born understanding quantum mechanics…and I'll show you a television character.

I say this to dispel a seemingly fantastic part of this history. Melissa went to school, and studied and studied, and then when she had some spare time she researched, which is to say — studied. It is true she was then adjusted by the AOU, but that is because she had placed her bowling pin on the alley of opportunity by, you guessed it, studying.

So, sometimes good things come pretty cheap. It does take an intervention by aliens, which is very rare. When it does happen, when aliens come a-visiting, they might pass things they learned in school on to a more primitive culture. Which really happened to Melissa. Melissa no longer considered it a gift; now she understood it better and was thinking it was more a rape than a gift; still, she determined to make the most of whatever it was. She turned those enhancements into some very sophisticated technologies over the metaphorical night. Imagine being a Super Geek and being shown paths of thinking you have never imagined. Either you get really depressed, or you jump on the train. We try to screen out depressives.

A really potent Super Geek Personality eats challenges for breakfast, right after an all-niter of challenge consumption. Depressive bright minds fall into a category of aggression that we made some effort to avoid.

Aggressive Minds — Two Kinds

There are two kinds of aggressive minds. Then, of course, there are three, four, five, etc. But, of the two kinds of interest here are the minds whose aggression comes from proving themselves to the world. The "I'm better than you are" mind, or the "I'm right and you are wrong!" mind. Another, and more interesting mind, is the "I'm better than I am" mind. It says "I'll show me!" This is not to say the outcome of any single comparative analysis performed upon these two types, where they shared equal intellectual capacity, could be predicted. Both will produce. But, one is a diminishing motivational force, over time. The other is sustainable.

"Beware giving power over your life to a person or agency; sooner or later they are bound to wield it."

—*Thodkin Marblemeister and/or others.*

Quickly, let's say you are of the "I'll show you" variety. Let's say you do...show me. I'm impressed...your motivation is gone. At best, you find someone else to set up as judge and do it all again. Or, I'm too stupid to understand your marvelousness, I shrug my shoulders...your motivation is gone. Motivation is a key to success. Self-motivating is a key to optimal personality. Show the world and you are done. Show yourself, and you can do it again, "I'll show me...Yeah, but can I do it again?" Having a *little self* challenging you is a much stronger force than a big world egging you on...over the long-term. We were thinking for the long-term. And besides, we had some people to show some stuff.

So, if you will try to remember that we — Melissa, the SGs, the law firm of LSLB LLC, and myself, not to mention a host of unapprised support players — all worked very hard to glue some basic infrastructure together over the next few months, I'll spare you a tremendous boredom of hearing about another human's mundane detail and just give you the high points. I'd sprinkle the story with our confrontations, but in truth, things went unbelievably well. This is not something you can expect, or should even hope for, although it is okay to believe things will go smoothly — all optimists do.

We Pillage Super-Geek Hangouts — Universities

We did some hiring. We did it discreetly and we did it quickly…the first part went this way.

Gold is where you find it.

- Buffalo University loses three SGs
  - electrical engineer
  - software engineer
  - chemical engineer from the nanotechnology department

We headed for MIT.

- MIT loses 17 SGs

A trip to New Mexico bears fruit.

- Sandia Labs loses three SG's
- Santa Fe Institute loses three SG's

I always wanted to attend Caltech.

- Caltech loses 11 SG's

I know people in Palo Alto.

- Stanford loses four SGs

The oceans are big places.

- Woods Hole loses three SGs

We then went after project people.

Meanwhile...

We were discreet, but we weren't unnoticed. Each individual university thought it had an odd run of bad luck. While some universities build in attrition numbers to their acceptance figures, prestigious universities count on their brights bringing recognition to the school. Losing your brightest is not a thing to advertise, so they kept it quiet. They each noticed, but they didn't notice the big picture. They didn't detect conspiracy. That was left to the Queen of Princeton, who, after some false starts, did get up to speed. She quite correctly decided that we would try to keep our knowledge to ourselves. And she did realize after some discreet discussion with experts that we would need both people and facilities. Incorrectly, she guessed we would end up at a university because she thought Melissa would go where she knows the culture…this left out corporate labs. Still, she was close enough. She was able to detect that the specials were missing from several universities she investigated. Of course, the barn door was wide open and the horses were loose on the prairie. But, it gave her hope. And she had what was, at least for her, a big picture to motivate her.

> CI_132 L0ARII:
>
> The Queen had hired an investigation firm. It had a good business hunting down Princeton dropouts for parents. And a nice side business of recording the activity of co-eds for worried parents from the Midwest. In fact, though, the firm was professional and well-connected in Princeton. And it had pieced together some events of the day in the lab. The report went something like this.
>
> A senior investigator stood in the office of the President. Thick wooden doors were closed and the recitation went forward in a monotone voice.
>
> "The deceased was recorded deceased three months prior to your meeting him. Cause of death was pneumonia. The

•••••••

* CI_132 Level 0, Archivist Guild Rivendell II. Created Intelligence responsible for archival detailing, verification, modeling and documentation. Note: All CI Archivist Guild members seem to share the characteristic of 'wanna-be screenwriter'. The success of each seems to reside in its 'modeling' software.'

afternoon of the death, the body became missing and was only recovered in your lab. An observation of excessive 'beer breath' by one of the campus police (specialists in beer breath) opened an avenue of investigation, leading to small success. Reports from The Frolicking Scholar Pub by several waiters who served the deceased over a time spanning several weeks said he always waited for the same booth. They said he was pale, though not ghostly. Once taking the booth, he occupied time by drinking many rounds of beer and tipping excessively to keep everybody happy. On the day of his second death, he gave a meal to a homeless person. They left the pub approximately the same time. Neither ever returned. A note about the booth indicates the jukebox selector was stolen the same day, though no damage was done to the booth and nobody noticed it until the next day."

"You are thinking the jukebox gadget is important?" asked the President.

"Not necessarily," came a noncommittal reply. "From your descriptions of the second male in the lab, as well as the descriptions given by the University Police, a high probability exists that the homeless person and the professor as you have identified him are one and the same. There were no visiting professors matching the description."

"The professor was a homeless person?" The President was incredulous.

"Most likely," was the only reply.

"So, Melissa, a graduate student at the most prestigious university in the nation, forgive my pride, was involved with a person who has been dead for three months and a homeless person impersonating a professor." She said this without inflection. "And she has run off with the homeless professor? I presume the dead man is still dead and accounted for?"

"Yes, he is, or his urn is. He was cremated. We found this regrettable, since we did not have the opportunity to examine the body. Few people die twice. It seemed important, though perhaps not to this investigation." It was difficult to tell if the agent had a sense of humor. From all outward appearance he did not. Still, perhaps wry wit was possible.

"We traced Melissa's luggage to an airline lost and found. The luggage was very new and none of the clothes had ever been worn. She has not picked it up and she didn't board any airplanes under her own name."

"Unlike popular dramatic media, our agents do not have access to commercial airport video surveillance, so we cannot determine if she boarded an airplane under an assumed name. Given excessive funding, such access may be achieved and confirmation one way or the other could be made — though even then it is not a certainty."

"She has not used her credit card and has made no withdrawals from her bank account. This begs a question regarding the source of new luggage and clothing."

"Her parents had not recently heard from her, but were not worried until we asked them questions. They graciously paid us a small fee to find her, allowing a termination of a missing persons investigation in this state. They didn't want Melissa thinking they disrespected her privacy. Melissa was apparently known to be headstrong and 'She's always been a little quirky.' I quote her mother. Subsequently, they dropped the investigation, saying they couldn't afford its continuance. But, they did not file their own missing persons report, and do not have financial difficulties. It is either that they do not have a close relationship with Melissa or Melissa has contacted them. It has not been possible to breach their transplanted Midwestern reticence. I suspect they have been contacted. I base this on the disparate level of anxiety upon first hearing Melissa was missing with comparison to the conversation terminating our contract. You must take this as inconclusive, yet supported by extensive agency experience.

"If we are correct in our assertion, Melissa is acting furtively, but has not been the victim of foul play by the homeless professor as you call him. As you might surmise, homeless individuals are very difficult to track. None of our attempts to bribe local homeless have yielded a single verifiable piece of information. Contrary to popular belief, homeless persons are often excellent sources of information. Observation is a key to survival when living on the streets. The homeless professor — though 'homeless professor' is likely an inaccuracy — has also disappeared." The agent paused, then looked up from his notes indicating he was ready to answer questions or receive further instructions.

"I have been thinking that Melissa might be looking for a job in industry or as a lab assistant in academia. Her work here was very good and she may have decided to take a sabbatical to make additional tuition/expense money. It has come to my attention that some discreet firm has recently begun staffing up a sophisticated project of some kind. And has managed to influence a variety of university students to join them. It may be that we will find Melissa with this firm." The President had heard of some of the losses at MIT, but she was inferring the recruiting agency. It was a good guess.

The Agent studied the President for a slow moment. "I surmise there is more to this investigation than you are willing to reveal. This may hinder our ability to assist you further. Do you have new instructions or should we conclude our efforts?" The agency chose college town work because it rarely, if ever, put the agents in harm's way. The agent would have been most happy to send a final invoice to the President, but could not risk alienating her. She did, after all, often put meat on the table. But, the agency was capable of detecting the trustworthy and the anti-trustworthy. The President was way too slippery to be trustworthy and quite probably leaned into anti-trustworthy.

"Would you be willing to undertake a second objective?" she began. "It has come to my attention that there have been a number of unexpected student departures from MIT. I would like to know if these can be considered coincidental or conspiratorial. While at the moment I do not know how they may connect to Melissa's departure, I cannot in good conscience allow one of my students to disappear without making every effort to explain the circumstances to the parents." She paused and gave a little lip-puckering, head-nodding look to the agent that he interpreted as *"Just look at how sincere I am, how caring. I'm lying, I don't give a damn about the parents, but I really want to know if these dropouts are in cahoots with Melissa."*

"You have a list or some documentation from which to begin?" The agent now became certain that something very unusual was, as they say, up. The President was totally indifferent about the kid's parents. She clearly had ulterior motives. Motives that she was not going to explain, but would continue to fund. "We will be forced to incur increased expenses. Shall we also increase applied resources?" This would solidify the conjecture.

"Please utilize whatever resources you have available. These parents deserve our best efforts." Again the pucker-lipped, head-nodding. Poker players and everybody who watched the classic movie *Maverick* know this to be an unmistakable, even blatant "tell." Whatever her words regarding her motivations, she was, for some reason, going into business for herself. She wanted to know where Melissa was and what happened at MIT. And that, as they say, was that.

"To whose attention shall I send our invoice?" The agent was unreadable and at the same time artful in his discretion. He offered her a path of payment which was not detectable on University accounts.

"Perhaps, as I think of it, the University should be allowed off the hook. After all, it cannot be held responsible for every child that becomes depressed and disappears." The agent noted yet another fabricated motivation. "But, I feel responsible since I was there at the moment of crisis and was unable to stanch the flow. Yes, please send further invoices to my address." Her puckered response added a glowing halo. Had the agency been less than above board, it might consider this an opportunity to extort higher fees. In historical fact, the agency wasn't above squeezing fiscal udders. But the agent felt there was plenty of time for that after the investigation revealed the true motivations of the client.

"Fine, I'll see to the arrangements. If there is nothing further?" The meeting was concluded.

The problem for unscrupulous conniving Queens is at some point their goals exceed their reaches. They must hire services or coerce services. Complexity grows almost instantly out of control. If lying wasn't enough, and we know it is very difficult to keep simple lies in order, she had just stimulated the development of multiple falsehood generation. And she had launched a set of completely autonomous individuals to stimulate ripples on the previously quiescent surface of the MIT campus. Quiescent in this context…there were plenty of ripples waving through MIT at all times, but up to now, the disappearance of 17 grad students was nearly without ripples. The agency would do its best to be discreet, but it was going to be difficult to be totally unobtrusive.

Fortunately, We Were Wired

We had a change model running now. Melissa and I were both applying cognitive resources to model the Queen and her possible actions.

We Made Our Way to Santa Fe

By this time Melissa and I were working smoothly together. We were having lunch in Santa Fe, New Mexico, about to be on our way after rifling Sandia Labs and the Santa Fe Institute.

"As much as you have exposed to me of your history, my own short life still prevents me from really understanding what it must be like to be so old…to have lived so long. How have you really survived 10,000 years of cultural change? It is inconceivable to me." Melissa was pretty hungry these days. Our modifications used much more energy than a normal body. She squeezed the question out between a steady stream of sopapillas filled with butter and honey. Waiters typically left our table shaking their heads — shaking a head; no waiters we had met to date had more than one. After our ordering they didn't seem to mind the request for a table for four. Presently, we had two stacks of sopapillas, chile rellenos, tamales, huevos rancheros, enchiladas and my lasting favorite, carnitas — and the omnipresent liter of draft beer. Our stomachs would be bulging before we could stop. It forced us to vary our cuisine regularly because we soon had eaten everything a culture had to offer.

"After a while, I became able to detect the impending changes heading for a culture. But my early years were very primitive. We had no models to follow and no history that could be trusted."

"I noticed some of the unbelievable imperatives you and your tribesmen lived by. As you remember them, do you feel silly? Or are you just amazed you lived through them?" She was slightly smiling as she needled me.

"I remember when I heard about the Anasazi. I heard they had vast cities built into walls of rock. I heard they had stores and warehouses filled with provisions for winter. That they had stone fences — corrals of fat desert animals. It was just like hearing about Atlantis, which we know is wishful thinking at best or a story designed to secure a dinner invitation more often."

Chaos by Thodkin

Attractors Draw Order from Chaos

Thodkin's Attractor Typing.

Non-chaotic Types
- Point
- Cycle
- Torus

Chaotic
- Strange

Attractor Strength
- Strong
- Meta Values
  - Excluding
  - Vibrating
  - Porous

Persistance/Longevity
- Annealed
- Soft
- Weak

Shape (Basin of Attraction)
- Narrow
- Shallow
- Wide
- Deep

"You told the story of Atlantis, to get dinner?"

"Oh, many times. Sometime, I'll get us a free dinner with just the true story of Atlantis.

"But as cultures became famous, like Incas and the Chinese, the tribes I spent time in began to evolve. Tiny ideas were copied. We began to create models of rumors. When the models failed, usually we had a quick political struggle and a change of power base, or the tribe dissolved all together. This happened many times to me, rarely more than once to my contemporaries."

"Did you meet the Anasazi?"

"No. I did try, but they lasted only about 1,700 years. As soon as I heard of them I began migrating east. I was in the Baltics at the time. I ultimately crossed the Aleutians and spent time as a member of most of the Pacific Coast tribes. My longevity continued to force me to leave one tribe after another. I was in Tlingit, Tsimshian, Kwakiutl and Salish tribes. Then I worked my way east. But I never went far enough south to get near the Anastazi."

"So that is how you got to Niagara Falls?"

"I've now been there three times. The first time I was making my way north to south, where I finally joined the Cherokee. I still think of myself as Cherokee; some cultures fit better than others."

"You walked the Trail of Tears?"

"No, I had to move on before it. The Trail was walked in '38-'39. I had gone back to Canada about 40 years earlier."

"1938? That wasn't very long ago."

"All science, not much on history are you? 1838. Then I headed back west. My European genetics allowed me to blend in with the rush of Anglo/Europeans thrusting into every nook and cranny of the continent. I began serious book learnin', the writing was on the wall. Literacy was going to cascade over the Americas. I decided to go as far away from the crowds as I could go."

"You just took off alone, headed west?"

"I did; the colonies were overpopulated. But it is amazing that even in sparsely populated areas you run into people. In fact, on that hike west I actually had a conversation with William Clark near the Tetons in 1805. I was on my way to San Francisco. I told him there was no navigable passage, but he said he didn't really care. He was kind of haughty about the whole thing for quite a while. He couldn't understand how I could travel alone and survive. And he didn't like that I could get along with the Indians without having to use tricks. But he liked having someone else around who spoke English without a mouthful of tobacco. I was trying to save him some hardship and was a little too insistent that he was being a fool. Finally in frustration he said. 'Look, I'm an adventurer…this is my adventure. Now, leave me alone.'"

"I guess someone could say that to us," Melissa commented quite dryly.

"Yes, and I would have a great quote all ready for them. I think, however, he was not speaking his true thoughts. He was an Optimist with a capital 'O'. Optimists can't afford to be realists. It would seriously debilitate them."

"As it would us. You know I don't believe you met him anyway?" Melissa was incredulous and a scientist. People have been making unsubstantiated claims to scientists throughout time. Scientists must disbelieve. So, I popped some beads and showed her.

"Thodkin, even though you are unbelievable, forgive me for doubting you." She swallowed another sopapilla nearly whole. "Boy, these things are good!" Melissa never wasted time on being wrong. Wrong was just another form of information to her. It got addressed, given an importance evaluation, factored into any immediate plans or processes, and discarded from her fore-thinking. Poof! It was digested and the waste eliminated.

"You wanted to talk about modeling the Queen?" I prompted her.

"Yes, I have some ideas." She mused and chewed a moment. "The problem isn't really how to model. We are constantly modeling all manner of things. The objective is to improve the model without really knowing her very well."

"I understand. I propose to you that the task is less difficult than it seems," I responded.

"Really?"

"Yes. If the Queen were you or I, it would be most difficult. We are unprecedented. Our abilities to partition and slant partial personalities allows us to innovate wildly from what might be anticipated. In essence, we can choose from a succession of coevolution mutations — cognitive Darwinian variation — and bang, we evolve. The Queen has only herself and any advisers she brings into the task. She can only bring members of contemporary society to the problem. And even the fringes of our society can be modeled."

"How? What tools can we use to define the problem space so that the model will yield speculations of higher probable accuracy than that of a politically astute mind?"

That was a well defined question, if a bit wordy, so I answered in contrast. "Movies, TV, the news, bestsellers and comic books."

"You are kidding? No, I see you are quite correct." Melissa never took long to catch the glimmer of viability in an idea. She didn't waste time discussing dead ends and never lamented them. She wriggled the value from a thought and left any lifeless skin behind. "You are, if I understand you, saying that the models that the Queen draws from are movies, TV news and popular artifacts. She is what she eats… intellectually. The whole world is a stage and she gets her lines from contemporary fiction and nonfiction, which is informed by fiction. And her culture only provides limited range of divergence. She will think within that range. You are undoubtedly correct." She paused a moment and added, "We must concern ourselves with this limitation of our own culture; I more than you."

"Yes, but we have an edge and we know of the tendency to draw from the imaginings of entertainers. Forewarned is…etc. To be precise, the Queen's culture is providing a narrowing range of diversity. As successful cultures age, they always become more conservative — limiting themselves. I don't mean we should underestimate her, but I think she will have to get very lucky or experience radical epiphany to do any real damage to us."

"Optimists can't afford to be realists. It would seriously debilitate them."
—Thodkin Marblemeister

"I would rather leave her without even luck." Melissa could be very determined.

"I do think there is something more we can do."

"You are about to suggest we develop a dossier of her parents, siblings and extended family in that order."

"Yes, you obviously are thinking along similar lines. Brains model. They model our parents, then siblings most thoroughly. Sometimes the order is reversed. To the degree the people we come in contact with during our lives are similar to our closest childhood relationships, we model them well, or not well. Everyone has met someone they have immediate kinship to. The very word expresses the nature of the meta-process. Familiarity is another word that expresses the state of knowing, of successfully modeling. Good models feel 'like family,' are

Main Entry: **fa·mil·iar**
Pronunciation: fa-'mil-yar
Function: noun
Date: 13th century
1 : a member of the household of a high official
2 : one that is familiar; especially : an intimate associate : COMPANION
3 : a spirit often embodied in an animal and held to attend and serve or guard a person
4 a : one who is well acquainted with something b : one who frequents a place

familiar, feel kinship. Something so obvious, but nearly totally out of consciousness."

"And the opposite is also true. We know who we don't model well… instant dislike," Melissa added. "Again language represents the state… dislike, not like, can't model. It is a good idea. I'll see if I can design some autonomous model signature codification agents to help us do this with some validity checking. Otherwise we could be responding to a model that is tainted and have no means for debugging it. We need genetic relationships and algorithms that point us more quickly to areas of fitness in the problem space." Now, that's what genius is. It is venturing out for accuracy in the midst of the overwhelming sand storm of chaos.

Rodders

As we talked, as we communicated on other wavelengths as we digested and ate (not necessarily in that order) I was observing the square outside the restaurant. A beautifully detailed '21 Model T hot rod drove into the square and parked. This is significant because the square had "No Parking" signs everywhere. There were no cars in it. People jaywalked every which way. The T drove almost 360 degrees around the square and parked on a precise diagonal. I didn't think too much of it except to imagine the pleasure of the fortuitous meter-maid/man bound to arrive shortly. But, this wasn't the future. Instead, in quick intervals other hot rods rumbled into the square and formed a perfect pattern of diagonalized vehicles around the square until, just in front of the restaurant where we ate, a yellow '32 two door sedan pulled in, followed by a red '34 with perfect flames spreading back from the rakishly slanted chrome grill of the classic full-fendered 1934 Ford coupe. The parade continued a little longer until the flow of hot rods trickled to a stop. The square was outlined in hot rods. Rodders exited their cars and milled around socializing or headed into the restaurants around the square. I noticed the yellow '32 sedan was driven by a woman in her early 30's who had a few words with the pudgy guy driving the red coupe. She walked over to the benches that ringed the monument in the center of the square, had a sip from the water fountain there, and settled down on one of the wrought iron benches.

She sought to enjoy a temporary liberation from the confines of her hot rod under the small shade of a recently planted sapling. The pudgy guy headed into our restaurant.

Melissa had finished off everything remaining on the table. "Had enough?" I asked.

"Plenty. I think I love Ms. Sopapilla. What a delicacy. You ready to go?"

"In a little bit. I'll get the check and I have to make a call."

She said, "I'll meet you outside," and walked through the hot rods to sit on the same bench as the driver of the yellow '32 sedan. She was carrying her backpack slung over one shoulder. We had hitchhiked the recent stretch into Santa Fe.

"You hitchhiking?" the Yellow Sedan asked.

"Yes. Hi, I'm Alice." Melissa responded. She had adopted Alice as her alias. She and the rabbit hole shared an affinity. "That yellow car yours?"

"Yes. You like it? I built it myself. I have always wanted to go on one of these hot rod caravan jaunts to the coast. I thought I should earn the right by building my own car. You like cars?"

"I used to race go-carts when I was a kid. But, I've had to leave them behind. Got a little busy."

"Hitchhiking take a lot of time?" The Yellow Sedan asked, with a quick smile. "I'm Aimee."

"It is slow sometimes, but gives me plenty of time to think. I've had a lot on my minds lately. What do you do besides show guys how a hot rod should be built?"

"I try to show guys how buildings should be built — I have a one-girl architecture firm. I've been studying and scrounging little home improvement jobs mostly. It can be difficult to convince people to think non-traditionally about architecture. It is most often an expression of their history."

"Like, this is where I've been? This is a show of my style roots? This is what I always wanted to be like?" Melissa took a shot at interpreting Aimee's remarks.

"Yes, that is pretty much it. Even when somebody decides to make their dream house, or a corporation wants to express an image, it is often built on something they have seen somewhere, that represents power or success. It's a shame really, because they never get originality, just derivation."

"Funny you should mention that. We were just talking about that over lunch," Melissa observed.

Three rules:

- Knowledge Combined with Superstition Equals Bleak Conservatism
- Knowledge Combined with Superstition Equals Power Overestimated
- Today's Superstition is Tomorrow's Standup Comedy

"Really, you were talking about derivation over lunch? Must have been fun. That is one of the reasons I like these rodders. Some are old school, trying to recapture their youth, but others are trying to build something both beautiful and innovative, non derivative but still genre bonded. Some are even ecology conscious." Aimee had become interested in the conversation. She was a little surprised to be having a multisyllabic conversation with somebody lugging a backpack. "Are you traveling with somebody? It must be a little dangerous wandering alone on the road with just thumb locomotion."

"Yes. I'm traveling with the guy talking to the hefty fellow by the red, flamed coupe over there."

In fact, she was watching yours truly doing a little more than talking. I was looking the car over very well. It was a work of art. I have loved '34 Fords ever since they came out but I had never owned one. And they didn't originally come with Pratt & Whitney turbines in them. Neither were they fueled by hydrogen. From the outside, you'd expect this to have a big V8, some monster carbs, maybe chrome pipes. But you would be surprised to find a concept car transformed like someone took an experimental vehicle and rubbed it vigorously with a magical classic '34 Ford detailing carnauba potion.

"The pudgy guy is a good friend of mine. He is actually the only person to help me on Girl. I'll tell you a secret if you promise not to tell anyone." Aimee spoke in mock conspiratorial tones.

"I promise," Melissa said, crossing her heart.

"The guy is wearing a fake stomach."

"Not really."

"Really. He is very health conscious, doesn't drink, works out. But, he loves the old rodders and wants them to think he is a beer drinking good ole boy. You should watch him nurse a beer — artful. He thinks they might ignore the bizarre things he did to the '34. I guess it has worked pretty well. But if word got out that he doesn't have a real beer belly he would probably be ostracized. Well, probably not, he'd just laugh it away." Aimee Yellow Sedan was very serious. Melissa was playing along. Girls are totally nondiscriminatory when an opportunity presents itself to disclose a male's foibles, or male foibles.

"Or osterized. You are not kidding?"

"No, I swear. He is a very peculiar and committed guy. Loves people of all kinds, makes everybody near him feel good and worthy."

"Sounds like you have a small thing for him?" Melissa prompted

"Yes, could be. But he has been more than a little gentlemanly towards me. Still, you never know with scientists. It takes them a long time to get down to feeling."

"I think you are right there. He is a scientist?"

"Yup. A real life genuine Rocket Scientist. A metallurgist really, specializing in high-temperature composites. You should take a look while the hood is up. It's quite a go-cart."

Melissa and her new friend were looking over at the rodder and me. I had the pudgy guy showing me the unbelievable sights under the hood of the '34. He was just telling me that he couldn't think of selling it. It was his pride and joy.

As usual Melissa was sizing things up. "Sounds like the ideal hookup for an innovative architect. So, what is the secret project you are working on?" Melissa took one of those leaps she is apt to do at any time.

"How did you know I was working on a secret project? Or were you talking about Harley?" Aimee was just a little unsettled at being understood so well by a hitchhiker. The truth about innovators is that they don't really want to be understood. They would much rather surprise people. They live on surprising people.

"I'm not talking about Harley. The not-pudgy pudgy guy? That was somewhat obvious. But I was thinking you are chafing a bit about being an architect." Melissa was very straight in her diagnosis.

"You are quite right on both observations. I wanted to design special places. Places that work for people instead of making people work for them. I have been designing software that evaluates an architectural plan. It applies a set of rules that will point out concept flaws in buildings and spaces. It can be employed in a variety of ways. You can yell 'fire' with it and watch where simulated people run. But it is bigger than that. It shows patterns of growth and use. Anyway, that is what I dream about on the road, now that Girl is finished. And I have lots of time to myself on this jaunt."

"What if you had funding?"

"I don't know anything about getting funding. I'm more of a worker. I know how to make things. I scare too easily to be a fund-raiser." Aimee shrugged a little self-deprecatory laugh.

"I know exactly what you mean. But, sometimes fortune smiles on us…no pun intended. What would you do if you did have funding?"

Aimee thought to herself that this was a very surreal moment. Something in the way Alice had stated the question sent a quick chill up her back. In less than an hour she was disclosing her innermost desires to a wandering backpacker. But, why stop now? Might as well add to the confusion. "I need a space, some very fast computers and a small prototyping manufacturing plant…and a whole lot of expertise that I don't have. I have this idea I call the IsoTown Project. I want to design whole towns that fit cultures, instead of forcing cultures to fit towns. I want them to be self-sustaining."

"Yes, it is becoming plain that the world is in a lot of trouble because cultures have lost the barriers that kept them apart. They used to be contentedly isolated. They could believe and act in any silly way they wanted to. But, the walls have all dissolved. Now they actually have to convince outsiders who have never been indoctrinated to accept untenable…tenets. Change once rode in on the hooves of the rare galloping Hoard — now it wings in day and night. You want to give them back their walls." To Aimee's surprise Melissa had stated a quick summation of Aimee's own primal motivation.

"It's never been done and needs knowledge I don't have. But I know architectural systems and there are some pretty smart people up at the Santa Fe Institute. Harley knows them all. If I had money, I'd bribe Harley into being my boy. Then I'd violate all rules of business and make him a Technological Partner in my little enterprise. He'd get the people we need. I may not be able to raise money, but I can sell. I can sell something I make. So, you asked. Sorry, I probably sound like a wild-eyed dreamer. But, it could be done. Buy your little club an IsoTown? Just add water and poof — safety!"

When you tap the core of people you get a gusher. And this gusher was fitting nicely into a strategic element of Melissa's long-term planning.

"I have been thinking that selling and fundraising are very similar. But, I know the distinction you are making. You feel you need to show a working product before you can convince rich purses it is a good idea, and before you can act with credibility." Melissa had been doing this her whole life. It was like wiggling cold toes into favorite slippers.

"Yup. That's me. When you come from an Asian family you are taught the value of hard work and proving oneself. You are always proving yourself to someone. First your parents, then every white person you come into contact with. Sorry, no offense intended. It isn't true that every white person needs proof. It is an indiscriminate compulsion on my part. If you are white, I'll show you!" Aimee suddenly realized she had revealed a deeply rooted weakness to a total stranger.

"I notice Harley is pretty white. Is that the deal with him?" Melissa just asked. She wasn't trying to advise or counsel. She just gathered information.

"I wish that were true. Because I have, slowly, learned to tell my compulsion from real hots for a guy. Harley is just a great guy. I forgive him for being white. But, it would be nice if he were more...Japanese. Maybe someday he will convert." She laughed.

"Here is the thing, Aimee. I realize that you are making a judgment based on the fact that I have hitched into town. And I'm guessing that you probably think you can tell me just about anything because we will never meet again. But, in these judgments you are somewhat misled by your presuppositions about hitchhikers. I am not blowing my own horn when I say I am more than I seem. And I will make a prediction that we will be running into each other..

While Melissa had been talking to Aimee, she had been keeping the occasional eye on me. I had concluded a negotiation with false pudge-belly Harley. We were just exiting the bank across the street. Melissa had learned that when I went into banks, things changed. So, she surmised that if something changed for Harley things might change for Aimee. And she was mulling over how to approach Aimee with regard to solving a certain problem she had been considering regarding a place to live, quietly out of harm's way, when Harley and I interrupted her.

"Hey," Harley said, directing it towards Aimee and sliding it to Melissa/Alice. "Say, Aimee, I find myself suddenly without a ride. I was wondering if maybe, for the next little bit anyway, I could ride in Girl?" Now, that was a fairly smooth bit of favor asking. And it would have gone off perfectly if Harley hadn't turned organic tomato red during delivery.

"I bought us a car," I said to Melissa. "And we've been permitted to join the Santa Fe Hot Rod Association on its annual Bob's Big Boy Carhop Service Burger Fetch. They are headed to Burbank, California. And I told them about Pecos Bill's, so they invited us along as semi-official Interim Oakie Pork Guides.

Melissa just shook her head. "Fine, can you drive a stick?" she asked me.

"You joke, but it's an automatic. Seems that Harley…Harley, meet Alice. Seems Harley thinks unassisted manual transmissions are over-deified. As are V8's. Harley likes hydrogen."

"Harley, have you quite lost your mind? You sold Subtle Force? That's his secret name for it. Might as well tell them." Aimee pointed the surprise at Harley.

"Well, he made me an argument I had a hard time refuting. And anyway, maybe we should finish introductions. Aimee Semple Kawasaki, Thodkin Marblemeister. I'm Harley Travis. And you are Alice?" He was very genuine as he hinted for Melissa to add a last name.

"Marchenland. Alice Marchenland. Hi. So, Harley, do you need a ride? Thodkin could maybe ride with Aimee, if you'd prefer to stay with…let me see, Subtle Force?" I had not seen this sneaky cute side of Melissa. She was doing a nice job of sending a single message to both Aimee and Harley. Now was the time for Harley and Kawasaki to form a merger. Jump in front of the bowling ball or it's going to roll out of town without you.

"Really, Harley, if you sold the Force, you are going to need consoling. You better ride with me." Aimee tossed Harley the keys to Girl. He said thanks and went to shift his luggage from Subtle Force to Girl. Just so.

"I guess I'd better get a bite. You've eaten?" Aimee was already walking toward Girl to collect her Harley.

"Yes, thanks. We need to make a couple stops before we leave town. See you later?" Melissa shooed Aimee away. Go see to Harley, we're fine.

"Later," Aimee called back and walked with some hip up to Harley.

"So, you got us a nice inconspicuous ride to L.A.?" Melissa laughed.

"Should do," I replied, making every effort not to give her the satisfaction of seeing me chagrined. "There is safety in numbers, goes the saying."

"So, I would have guessed that he would never sell that car. How did you do it?"

"I offered him an unbelievable amount of money. But he said no, quite without a qualm. So, we just chatted. He couldn't stop looking at the '32 sedan. Finally, I suggested he might have a better time on the trip, without a car." I suppressed a Cheshire grin.

"What is impossible is impossible…of course, the opposite is also true."
—*Thodkin Marblemeister*

"Just so," Melissa nodded in knowing agreement. We put the packs in the perfectly finished trunk of Subtle Force. "I need to make a few stops before we get on the road." She said this after taking a quick but thorough look at the engine compartment, the undercarriage and a contorted look under the perfect dashboard, committing all to photo memory. The instrumentation is state of the art and the turbine has very few hours on it, almost new. The top is not chopped, but the car it is channeled…plenty of headroom, but still a little sleeker than factory.

After a couple of stops, the purpose of which I had not deduced after a few hours, we joined the auto-train at the entrance to Interstate 25.

"What is Pecos Bill's?" Melissa finally asked after we were through the logistics of getting under way.

"It is a little, very long-lived sandwich joint in Glendale, just on the cusp of Burbank, California. They make sliced beef and shredded beef and pork sandwiches that are just perfect. Order with light sauce and a touch of special sauce. I'll buy you a few when we get there. It turns out these hot rodders love their food. And besides, it is just east of Bob's Big Boy."

"You've been to Bob's Big Boy?" Melissa is not really surprised.

"Many times. Try ordering a Big Boy with fries, hamburger rare with heavy relish and a side of blue for dipping burger wedges and fries. Nothing better." I had a real hankering. "Also, Burbank is spitting distance to Pasadena."

"Caltech," Melissa affirmed understanding.

"Yup." We traveled in silence with our thoughts for a while. The car had a smooth silent ride and was gentle but muscular when asked to be. It was a real find. I did the driving on this first leg, while Melissa worked with her hands. She had purchased some computer components, various cameras, black boxes, a battery-powered soldering iron, a GPS unit, some mechanical gadgets, a small welding torch and a laptop computer with extra batteries; also a rather large and fairly heavy cardboard box with the Mercedes-Benz logo on it was wedged in the last of the available space. At least she wasn't asking "are we there yet" all the time. I cruised with the stereo cycling through a professional quality disk changer. Melissa bought gadgets, I filled the CD changer. Melissa chuckled when it hit the ZZ Top Eliminator album. Who would have guessed a Princeton geek to be a ZZ Top fan?

We stopped for the night and managed to secure a room at a motel nearby the Santa Fe rodders. We had to run an extension cord out the motel window to power the electrolyzers that recharged the hydrogen cylinders. They were built into every nook and cranny of the car. For those of you out of touch with hydrogen, there are ways to bind it then release it in a controlled way. While it is bound, it is quite safe.

Locals recommended an all-you-can-eat truck stop and we had another feeding. Then I got some sleep. Sometime later I thought I heard a rumbling car join the turbine sound of Force, but it went away quickly so I ignored it. I don't know if Melissa did much sleeping.

The next morning she took over the driving responsibilities. This is not quite accurate. It seems that with some parts from a new Mercedes, and the aid of a local 'old school' hot rod builder's pristine garage, she and her fancy hyper-brain designed and implemented an Auto Driver — overnight, while care-taking the Force's beauty and excellence. Think of it as the ultimate cruise control. And it worked… first time. Those of you who know about such things will consider that to be a big lie. But then you don't have Melissa's brain; of which, in fact, the car did have a little piece. She had done all the programming in her own brain, then dumped it into a set of micro-controllers in the gadget she had been making on the previous day's drive. So, she was driving hands free. It was somewhat unsettling.

"Very impressive!" I remarked with enthusiasm. "Maybe you should put a hand close to the wheel so we are not too conspicuous. Or we could buy a fake hand and just hang it from the top of the wheel."

But Suppose Everything Works

Melissa took a little breath and allowed herself a rare gentle sigh. "The thing is, I have just had a most interesting night.

"It began as a means of convincing Aimee Semple Kawasaki that we are not trifles. She only understands proof. This should be something quite outside her ability to doubt…or believe. And I am counting on Harley to reinforce an assessment of the accomplishment's relative difficulty. Well, I did it for that and I thought it would be kind of fun to get dirty."

"But, I was working in a totally new way."

She was going to get it all out, so I just watched the road, tried to ignore the fact that nobody was driving this 1000 horse power turbine, and waited.

"You know I have read a lot of science fiction — mostly before middle school. But I read then voraciously — old stuff, new stuff, antique stuff. And there are certain things that are almost universally glossed over."

"Like, if we figure out how to make everybody smart, how does that actually work? And how does it happen to you. And if it does, how do you use it? And if it happens to anyone or everyone, how does that work. And can it really happen all at once and 'poof' you're capable beyond your dreams? And, really, who cares? Why do it? Why not just live as we are and croak when it is our time? And if it did work — well we've talked about RIFT, but I didn't feel it where I was last night.

Last night I wasn't doing, bio/nano anything. I was making hot rod. I know nothing about hot rod, really. But, I did it anyway. Or 'They' and I did it anyway."

"Last night, it was like I was flying in a dream. It was effortless. It was joyful. There was only one me, but I had great helpers. Oddly, the 'me' part of me was doing all the grunt work. I had to drill and weld and machine and wire and mate up parts to other parts and paint all the physical pieces, while my other minds informed me of dimensions and material requirements and the color of the wire that went here or there and how this or that was supposed to be done."

"So, cool, my night was great! It was amazing.

"But, all of that sci-fi, it rarely dealt with the 'How one gets smart'. The most common was radiation poisoning. Everybody gets a big head like *Illya Kuryakin* in *The Outer Limits* and they play Mozart with six fingers. Or a great idea, one I thought could actually work, was knowledge carrying bacteria. Or progressive brain replacements in *Nanotime*, that is sort of what the AOU did to us only not little chunks — atomic chunks. Or there is some genetic mutation, disease or pill. Or you come on the action later in the game like *Egan's Diaspora*. It is great and visionary, but it pretty much starts with everybody already different — whole cloth, no weaving of the new patterns. And one thing is certain, nobody goes to school to learn anything — knowledge just happens. It is the Doogie Howsering of our future."

As she took a breath, I thought it was a good thing this road was so long.

"So, now I know what it would be like to be infected with knowledge bacteria, or irradiated into another appendage and oversized cranium. It is, and I don't use this word lightly, awesome."

"What is more, I'm becoming very optimistic that this is where the species is headed in the near term. We are going to make this happen. Everybody is going to get pretty damn smart, pretty damn soon, and pretty much all at the same time — what then?"

And so, that was it. That was where she landed after her flight of joy.

And so I said, "To state your enlightened destination in the positive..."

She blinked...in the positive. "How will we make a place for all of that amazing and joyful potential!"

That was it. That was the task. She had stated it in the positive. That is what she had committed to. Not making the next big thing — the 8th wonder of the world — though it would be. Not wringing her hands about handbaskets to the brim with hell, or talking in circles while sitting with humanity's demons. But, she was just going to have to make room for a ton of potent joyful intelligence.

I turned to her, gave her a little smile and a slight quick nod. "Yup."

I took my own little breath and gentle sigh. After a little while we got on with it.

"You have a plan for Aimee?" I asked, knowing she did.

"I think she can forge the tool for us to construct a place, a town from which to extend our goals. We can't be buying a trail of hot rods, at least over the long-term, as much as I do like your choices and innovation in covering our tracks. I have never had so much fun. But, we are collecting people. I predict that within the next six years, we will need a working community and infrastructure to support all of our endeavors. Call it a campus. And, I think it may turn out to be quite useful for it to be self-sustaining. It seems likely that somebody is going to want to disconnect our power. I like Aimee's tagline...Just Add Water. We need an uninterruptible supply of safe water. Other needs will flow from achieving that."

"You are surely right. In fact, it might be prudent to develop more than one such place. So, we must have a discussion with the Kawasaki/Harley merger."

Melissa asked me if it would be possible to fund initial IsoTown research. I had paid a small fortune for the Subtle Force. Harley could fund the start-up investigation of just about anything. Then we could come back in and supply second round financing. The initial period would be a good test of the IsoTown management team. We were in agreement with this plan. Melissa added Aimee and Harley to our watch list. Somewhere 1,000 miles away an inconspicuous lab responded by spawning dossier agents to monitor all media channels for information relating in any way to Kawasaki/Harley. So much for privacy. I once found a friend of mine by discovering a local newspaper coverage of his son in a high school basketball contest...a son I didn't know he had in a Midwest town with no gas station that I didn't know existed – the gas station didn't, the son and the town did. You would be surprised how many actions lead to publicly accessible information. One had to be most discreet to be hidden in this age.

Driving America

I would like to return for a moment to Melissa's very first analysis of the effects of the changes she was able to bring about in a living human. We had some time to discuss this as we cruised I-25 to 550, then west on the 64 and jogged north to check in on Four Corners, the only place where four state borders come together in one nexus (most of you knew that, but for those who didn't...).

Think for a moment about living 1,000 years. You probably will find that difficult. Think then about what makes you want to live. Most of you might not even have an answer to that question. There are cultures where you would be laughed at for asking the question. You don't really question living, you just live. But some of you need reasons. You have lived through strife. You keep doing it because...you are depended upon, or think it is a sin not to, or just don't know what else to do, or think that sooner or later it will get better. And for most, it does get better, or it at least becomes tolerable. What motivates you to live? It is a scary question, because the answer is often more a rote answer than a whole-being truth. We can tell the difference.

"People generally live as long as their bodies keep sustaining life.
That is all going to change."

—Thodkin Marblemeister,
The ~~Prophet~~ Patient?

Now, really think of your answer. What motivates you to live? What do you live for? Probably, whatever the answer, it will get you through 70-90 years of living.

Almost any reason does. Certainly until now, people generally live as long as their bodies keep sustaining life. So, that is nearly 100 years. What if it were 10 of those lifetimes? And if that is possible, and it surely is, it will most likely be unnecessary to die at all. It will become choice, not physiology. Are you ready for that? Can you generate enough joy or interest to motivate your living 1,000 years? How is that going to be possible?

The truth is that all the reasons people give for not extending life — playing God, against natural law, unsustainable (over-population) — are examples of the hindbrain trying to avoid addressing an unnamed fear. It is the fear that there is no *reason* in living. For large numbers of people in modern countries, the good life has become an arduous job — where death is the weekend off. But, that is not how it has to be. There are millions of people who look forward to every day. There are people who live through trouble and even in the trouble want more life. What is it in the First Minds of these people that produces this charge for living?

More truth. If we are ever to understand the universe, and I mean knowing through personal experience, not giving in, not traveling the coward's way of noncognition, not blind belief. If we are interested in piercing the big secret being kept by the universe itself, we are going to have to live longer. So the changes cannot only be to longevity, they must be also to how we think. To how we approach living. To how we find joy and motivation in life.

Melissa said we will need modifications to enable us to discern finer-grain pleasures. To be happy about smaller, more subtle things. To design happiness refinement thinking styles. Of course this is risky. Some people will fall by the wayside and not discover the origins of the universe. They will fall prey to some recursive pleasure. A pleasure that is self-locking and discriminatory, like drugs and television and power hunger, only hugely more powerful and constraining. These people will be casualties along the path to truth. But I am not too disturbed by those who fall into these choices, if they are choices. I am reminded of the seekers of the Eyes of the Overworld. Through these eyes all things appear opulent, all food is delicious and all people are intelligent, witty and fascinating. If people make the choice to bind their First Minds in the appearance of truth, so be it. Our goal cannot be to prevent those choices. Our goal must include ensuring that opting out of the future is only one choice and that most minds would have the opportunity to choose…unblindered.

Pecos Bill Runs Out of Sliced Beef

Somewhere between the Four Corners and Pecos Bill's, Kawasaki/ Harley had made interesting plans. A few days later, Pecos Bill ran out of sliced beef. I should have called ahead to warn them I was trailing hungry machine artists. And then the nearly extinct New Mexico breed of Wheeled Guzzlers invaded Bob's on Friday afternoon. The rest became history.

Melissa and Thodkin have Lunch Conversation with Harley and Aimee

Bob's Big Boy Restaurant in Burbank, California, is near — some say within — a community called Toluca Lake, home of the late Bob Hope. One section of the restaurant forms a little alcove. The alcove has one muralled wall and one with tall windows. If you get the right booth, you can sit there and observe people awed by cars parked in the carhop portion of the parking lot. Melissa, Aimee, Harley and I sat there. Everybody had burgers ordered just right — rare, heavy relish, side of Blue for dipping. Melissa and I were on our third, while Aimee and Harley nursed their first.

The Sante Fe contingent was somewhat impressed by Melissa's proposal of stimulating the launch of Aimee's IsoTown Project, but not enough. With no words of warning, no heads up, Melissa reached out and placed her pinky in a glass of ice water residing in perfect theatrical position center stage of the booth's table. Barely visible, little fila extended from it…soon the ice melted, and the water began to boil.

I offered explanation. "Melissa has decided that talking will not convince you to move forward. She cannot disclose all that she would like. That is for her protection, but also for the IsoTown partnership. She is undertaking another motivational strategy in the form of a demonstration. What she is doing is tasking computational nodes comprised of nanoprobes resident in her finger with a part of a problem she has been working on. In order to prevent her finger from burning up, the nanosites have created a very efficient heatsink from buckytubes that are now blowing off steam in the water glass.

"To sum up. Without so many words, indeed without any words, Melissa is telling you that she has more brains in her pinky than anybody has in their whole head. She is hoping this will convince you to focus on creating IsoTown, Inc. And she hopes you will consider incorporating some of her materials research — that has just boiled away her water."

Melissa moved her pinky into another water glass and chimed in: "Thodkin is sometimes a dramatist. In this case he understates. There is nothing like Thodkin or myself anywhere else in the world. We have agreed to give you an impossibly rare choice. You have highly desirable abilities and you have personalities unfettered by most restraints. We consider our meeting you wildly and unexpectedly fortunate. I also find you very enjoyable company, but I must ignore that for the most part. Nice people confuse me. I'm only now learning to discern the genuine from the false. I rely on Thodkin for judgment of resilient integrity in others. I would never have undertaken this little exhibition without his assurance that you would not be adversely disposed by it. I have mentioned that we will soon be needing facilities. They will represent a challenge far beyond the norm. If you find yourselves interested, we can provide new, patentable inventions to forward the IsoTown project. These inventions are directed at our own goals, but could easily be the foundation of a vast fortune for the IsoTown project, supplementing your goals. You could think of us as your first customer, a customer combining technology partnership with an early source of income. You would have cash flow, the entrepreneur's golden egg. We will negotiate technology licenses when required."

Harley spoke first. "We are unfettered, but you are kind of tossing us a harness and saying 'put this on, mule.' I seek to detect a fine distinction. It is possible to construe your disclosure as an expression of respect."

Aimee was direct. "I have seen three things that impress me, in no hierarchical order. First, a hitchhiker buys Subtle Force for a small fortune — a feat I would not have considered possible, no matter the size of the fortune. And I still have only a cryptic explanation of that happening. Second, another hitchhiker modifies Subtle Force to drive itself…in a single night — something I know to be impossible, but supported undeniably by the 'Addam's Family Hand' always in the front seat. Third, a seemingly human finger boils water in front of my eyes — and not a big index finger, a baby finger, a pinky!"

Aimee had become a little giddy. She paused, looked up and away from the table to choose her next words more carefully. Her body was shaking as she tried to force it into relaxation but only achieved an oscillation she was struggling to keep from her voice.

"During the last few days, Harley and I have conjectured often about the new owner/owners of the Force. In the end, we remained largely confused. This latest display serves only to fuel the confusion while raising the stakes. How is it possible for a human to boil water? What are nanoprobes…and bucky tubes? How did you get them in your finger? How did you tell them to boil the water? What were they calculating at such intensity to get so hot? Did you get the answer? Also, why has Alice disguised her name? Harley speaks German, allowing us to doubt your real name is Alice Marchenland (Wonderland). The name Thodkin, we have decided, is probably real; who would invent it? And lastly, if this is really possible, that you have more brains in your pinky than I have in my whole head, how has it affected you? I can tell you that it affects me — I am both frightened and excited. I have just stumbled onto the set of the *X Files* — or maybe *Candid Camera*." She had a good memory; she got it all.

Harley supplied the unspoken issue. "And, the final question after the last question is — if you have not been harmed by whatever you have done to yourself, and you are indeed still largely human (this may not be relevant), how do we get it done to us?"

All good questions.

Discussions with Primitives?

Aimee spoke up again before answers came forth. "So, do we seem like primitives to you? How do you cope with the slowness of our thinking if you can process information with such capacity?

"It could be an annoyance, but it isn't. It would be if I designed my inner experiments — my partials — to imagine that the way it works has alternatives. I just don't design self-destructive conflict into my inner multi-selves, except in double-blind experiments that never manifest in outward behavior. Each self/partial is working toward the same end. Each has the goal of producing optimum results; results that are sensory specific, that can manifest in physics, or are intellectual tools for affecting the environment. There are ways to speed up the organics, and there are ways to adjust to change more quickly and with less conflict, but when change is too fast, it is too fast. So, I make the input to my organics slow down. I am beginning to represent an aspect of personality that can be characterized by the question, 'What is the hurry?' If things go according to plan, I will never cease to exist — in one form or another. If I say the words 'life goes on,' I mean it about the individual, not some ideation of species affiliation. Something Thodkin calls the First Mind will go on and the relationships of knowing will go on; principles — and some factual information — may hole up in perpetuity, though unlikely.

"The means of creating my mind is evolving, meaning my tech is changing, is all. The idea of 'species' will soon only be a very distant point of origin, as remote in nature as our missing link, or the first cellular mitosis." Melissa was rolling now. She had been learning at a rate she was barely able to cope with and this was an opportunity to offer some of it up...sort of a peer review.

"I don't write the conflict of a movie script into my learning experiments. That is done in movies so people will recognize the conflicts of their own lives in the works of art or capital they confront. But where do those conflicts come from? Most are self-made. The future equals the present if every day is an uncontested version of its predecessor. Instead, I seek the beauty of resolution, of solution, of understanding the big picture. The observation that people create problems so they can avoid bigger problems does not apply to my way of life. If anything, it is the reverse. I am shucking the common human dilemma theme in search of kernels capable of growing past it. There are other themes to be discovered once we break the stranglehold of primitivism and time." She paused to review the issues and decided she must also explain a measure of her own limits.

"The future equals the present if every day is an uncontested version of its predecessor."
—*Melissa*

"And as yet, I haven't learned how to use my enhancements. There is wasted capacity because my personality is old school. It can't just become something else overnight. My brain is organic. And if I have the potential to be something 1,000 years into the future, I am not it now because I don't know what the new school can be. It is a bootstrapping problem. I've got ingredients, what I need now is a recipe. And then, after all that is said, if I become dissociated from my species, whom do I play with?"

I thought that was funny. Melissa claiming to be old school. Such a kidder. And Aimee…it amazed me how radical, when confronted by bizarre, suddenly turns into conservative. But it was really for the better; it meant that they were taking the possibilities seriously. It is quite easy to say, "I thought of that years ago," after somebody else has done the grunt work of actually achieving; it is very much another matter to take up the objective early and commit to it. It is very difficult to be ahead of one's time, because each moment in time has vast, powerful attractors attempting to assert and reassert themselves. You might say, nothing succeeds like a successful attractor. Evolution is riddled with overly strong attractors, riding saddleless into the future on the limits of some species or another. We are awed by the unchanged longevity of the tortoise and the alligator. Still, evolution played a mean trick on these species. It developed a system so strong that it could weather the changes of everything going on around it — without the need to change one little bit. Evolution never plays with success. Old school? Go to alligator school to discover its meaning. Alligators are simply annealed attractor orbits. Evolution is the archetype "away from" motivator. The moment a mutation results in assured survival, that is the moment evolution slows to a stop. It can take a chaotic avalanche to kick start it again for a species.

Space in Your Thoughts

"Maybe you have something different in mind for IsoTown than I do." Aimee began the disclosure process. "I have been focused on the nature of living space. My philosophy is evolved from the premise that the spaces we inhabit are not separate from us. We are, our minds are, inclusive of the physical spaces we exist through. For example, I know somebody who pays rent on an artist loft but almost never goes into it. He does this because the loft is crafted to stimulate him in certain ways. Once it was created and after living in it, after it actually worked as a space that helped induce creativity, he became able to visit it in his mind. When he did, much of the invigorating effect was experienced.

But something about it actually 'being' in existence in the real world is necessary to develop the state of mind. And it is worth expending tangibles to maintain its physical existence. There is always the potential for a moment within the loft's walls — a potential for silence, seclusion and exploration of irregular concepts." She was saying we run models of space in our minds, not just people and processes. I pondered this while she spoke on.

"Another smaller example is a person I know who is uncomfortable at parties — social gatherings with no defined objective but to gather. Or maybe the objective is really sex, or less optimally the shadow sex of social intercourse. He is uncomfortable until he finds the piano bench. If there is an available piano bench, anxiety levels drop and he is able to participate, even stimulate a communication that produces value for time spent.

"There really is no way to separate the physical world from the minds of inhabitants. Habitat is an extremity of the inhabitants." She was calming down, now that she was back in her world. "Girl — my car — is such a place to me. Perhaps Subtle Force is like the loft to Harley. Maybe as long as it is in good hands it is still part of his habitat. I have become sensitized to this idea. It seems part of a set of ideas that produce wild variance in emotional experience but are nearly completely unobserved. Except feng shui, of course. I could state my goal this way — I want to quantify feng shui and build it into habitats. No hit-miss, no superstition, no 'newage' con-job. Do it, always."

So, Aimee was nearly recovered. And she was right about Harley. She just hadn't fully discovered that her model was also part of his habitat. A much more powerful part than his car. I added two cents. "It is clear that we have torn down the safe havens of many cultures. The invasion of tourists…of transmitted images, magazines, free network access, goods….these all have played roles in dissolving the sensorium's perception of protected habitat."

"The issue extends to culture, heritage." Melissa took up the exploration. "We have seen a rise in behavioral extremes — ter-

[ADD]
Attention Distortion Disorder

An imbalance between the concern over who is right versus what is right.

rorism, radical conservatism, reasserted proselytizing by dogmatists. Maybe this goes without notice as a trend. But it correlates nicely with the destruction of borders — physical, cultural, lingual — and the incursion of process changes that include concepts like societal pace. As the traditional pace of a culture is altered it gives rise to what may be called a collective version of RIFT. People literally look around and don't see home, and they don't feel at home.

"Naturally evolving diversity is diminished when all roads lead to anywhere. Meme infections — epidemiology studies this — spread via contact. Strong memes destroy the weak, reducing diversity and increasing eco-fragility. The strongest barrier surrounding a culture remains language. But the adoption of common communication forces common language so the barriers again disintegrate. Unorthodox ideas get translated, oxidizing a culture's norms. Coevolution or merging brings about rapid evolution for a while, but ultimately leads to stable ecologies. Stable ecologies can last for a long time when outside variations are minimal. But interdependence can produce a house of cards, avalanches of destruction towards extinction, accelerated entropy or social decrepitude.

"In the end, partitioning, at least the ability to partition, improves viability. This is true for many systems. How do we discover optimum levels of adaptation and resistance to adaptation? How do different domains interact to enhance or diminish a domain? We must study." Melissa ran down.

Conversation stopped as a waitress asked if we had enough hamburgers. I bought lunch and we went out to open the hood of Girl and Subtle Force. Night was falling and the early gawkers were beginning to mill around Car Hop Alley. There was time for a little sinking in; IsoTown wasn't built in a day — well, the first one wouldn't be.

Somewhere Investigators Bug Ex-students

CI_132 L0ARII:

NanoTs, one of the first of the Toolshops of ISUS, was visited by two men in Hawaiian shirts. The nanoTs facility wasn't much.

It was a start-up and inhabited a small building in a tech incubator in Virginia. It had one big open space with a small glass room positioned in the very center. Computer workstations and a variety of esoteric lab equipment were placed in seeming disarray on a floor of unknown material so white and so clean nobody would hesitate to eat off of it. If an observer were very perceptive he might notice that all workstations had a view of the glass room. The glass room itself was completely empty at the moment.

There was only one exception to the sense of total openness of the facility. That was the reception area. It was quite small and had a floor-to-ceiling wall made of stone and an ordinary door with the obligatory cardkey swipe strip to prevent access by those without permission. Cardkeys were purely cosmetic; other means were being employed to validate those authorized entry past the stone wall. The wall performed the same function as a privacy panel on a desk. It kept unwanted advances at bay. The non-Hawaiians represented an investment group looking for opportunities to ground-floor nanotechnology. Of course, they were private investigators following up on the Queen of Princeton's list of college dropouts. Several of those on the list were on the other side of the stone wall and had the secret passwords and gestures allowing admission. Geeks seem to enjoy faux intrigue. In this case, passwords and secret gestures were just for fun; more sophisticated measures were at play.

"Mr. Nonamoto will be with you in just a moment, gentlemen. Can I get you coffee or a soft drink?" Perfect delivery of this line belied two things. First Mr. Nonamoto was a cue that bad guys were in the lobby. Second, nanoTs did not employ a receptionist. The receptionist doubled as Chief of Security for the little group. In fact, she was a physicist working on a new breed of sensors. The console she sat behind showed readouts from a variety of systems arrayed around the reception area. The simplest of these behaved like a tuned antenna array. They shifted frequencies to detect everything from materials, to objects, to fluctuations in heart rate, to air pollutants.

Eventually, Mr. Nonamoto was found to be too busy to see the visitors and one of the ex-grads would substitute. Mr. Nonamoto did not exist, but he did have every i dotted in the human resources filing cabinet. A cabinet of low quality, locked with a suitcase padlock, as a joke. Something anybody worth the title of private investigator could open…blindfolded. Or a kindergartner…blindfolded. While Mr. Nonamoto attempted to manifest in reality, the visitors were forced to spend idle time. It was useful time for the nanoTs team. Every free person behind the privacy curtain would be exploring the sensor data, and the elected substitute would be fitted with a deep ear transceiver to allow information to be shared during the incursion.

The substitute exec showed the visitors into the secure area after ensuring they signed in and got a nice paper name tag stuck to their flowers. The tour took little time. He answered the one interesting question regarding the custom stainless steel vacuum chamber with, "Sorry, I can't talk about that." There was little to see and nothing to demonstrate. The investigators explained that it looked like the team's investigations were too long-term for their New Venture Group, thanked the former student-cum-start-up executive for his time, and departed without remembering to offer the bogus business card the agency had invented. The speed of the tour didn't prevent them from hiding several illegal listening devices throughout the lab. Nor did brevity prevent the Chief of Security from sprinkling the non-Hawaiians with devices of her own so sophisticated that they would be mistaken for paper shavings without a microscope, and indeed would biodegrade very quickly when instructed after being interrogated by other sensors. While inside on the tour, the agent's rental car was sprayed with a clear smart paint by a kid passing by on a skateboard — a former grad-student. The car became able to broadcast conversations several hundred yards and the tracking relay drone would bounce signals anywhere in the city.

The notes developed by the agents would document the little girl sitting on the floor of the glass box. They would describe the drawing pad and crayons on the floor beside her. They did not find this unusual, just barely noteworthy. Apparently they considered the glass room a daycare solution for one of the employees. No explanation was given of the girl or the box. It was ignored with perfect congruity by all nanoTs members in the facility. The agents, trained to detect anxiety or avoidance observed nothing to make them attend the issue of the glass box and therefore missed the only useful clue. Several nights later, they would attempt and succeed in surreptitious examination of the contents of the filing cabinet. They would not note the fact that the coloring book was still in the glass box. They would discover the hiring dates of several of the dropouts specified as investigation objectives. But, there was nothing more substantial than an unusual string of coincidences remaining after they had exhausted their resources and performed to their skill levels.

Howard Johnson Hosts the Queen

The President asked to have the briefing at a Howard Johnson's Coffee Shop just after the breakfast rush. She sat in the booth with two suited agents, one of whom was the same agent who had delivered the earlier report on Melissa's disappearance.

"What have you to tell me?" The President was a little abrupt, no doubt for more than one reason.

"We have infiltrated a start-up company that was the destination of several of the students. A list of students found is in the document before you. The company is funded, but minimally, and no evidence of second-round funding has been discovered. The employees appear happy and eccentric."

"No doubt, a helpful bit of intelligence." The Queen seemed to barely control her agitation.

"Just so. As a measure of relative coercion or conflict over dropping out of school the subjective evaluation has significance." The agent was used to the value of details being misapprehended by his clients. "Security of the physical premises is lax and poorly maintained; for example, cardkey locks were disabled. However, network and computer security is superlative. We have been unable to breach nanoTs's firewall and no success has been achieved at investigating what we call 'loose-lip' sources. No communication of any kind has left any trail leading to the actual nature of their research. We are assuming their stated purposes are fabrications. This, at least on the evidence, seems an error in judgment. With the exception of the lack of information stemming from any source outside of the nanoTs building itself, we would conclude they are working on what they were incorporated to work on — advancements in nanotechnological fabrication technologies."

"Bullshit!" The Queen had fallen from reason. She existed now on a stark plane of compulsive/obsessiveness. It became clear to the agency that no résumé enhancing work was possible here. It might even become difficult to collect due payment. Payment schedules would have to be revised to C.O.D. before further work would continue.

"Assuming your motivation to undertake this investigation was to turn every stone, we investigated nanoTs." The agent barely held a smile from his face. He had found it bad business policy to laugh in the face of a compulsive/obsessive. It always resulted in more risk for less value as the compulsive reacted to the fear that they were indeed off the deep end. "NanoTs was funded by seven investment groups. These groups are members of an investment consortium. The venture capital investments represent the 'extremely high risk' component to a conservative but carefully considered investment strategic plan. The plan allows for a small amount of money to be placed in potentially high yield but low probability opportunity…"

"Yes, high potential, high risk. I understand the principle. Who funded the damn start-up?" Bad to worse, the Queen kept hearing failure in every new disclosure.

". . . to help the youth of our nation." The agent ignored the escalation of intensity. "NanoTs is funded by the Aunties and Grams Bucks for World Traveling and Better Christmas Presents, LLC. It has no helpful acronym." The agent achieved the professional monotone only by employing extreme self-discipline. "The organization, under different names, has existed loosely for over 200 years and has grown from a single crocheting coffee klatch into a widespread syndicate…of crocheting coffee klatches. They are consistently successful. So much so that they acquired a small but diligent travel agency to manage tours and accommodations for their membership. They were very forthcoming in explaining their investment strategies, inviting us to join in. Apparently, there is no gender bias regarding Aunties and Grams." Again the agent resisted a surge of very rare giggles.

"And the founder of nanoTs? Who did they fund, for God's sake?" She had lost hope of useful news; this was more for form than the expectation of operational intelligence.

"Lee Nonamoto, a naturalized U.S. citizen. He became a citizen shortly after the end of the second world war. He was born Japanese, though the only evidence of that is in his immigration and naturalization documents. They indicate he is nearly 75 years old. We did not meet him, but his name is on the incorporation papers. Our guide implied Nonamoto to be somewhat reclusive, but said he spoke English perfectly and was 'a very cool guy' who pretty much left the employees alone. We have no other information with regard to Mr. Nonamoto. He has an expired driver's license on record in the state of Washington. It is possible he has been out of the country for some time since there are no Nonamoto references in any source we have discovered that do not pertain to the founding of nanoTs. It is possible he is the missing 'professor.' Was the professor of Japanese decent?"

"I wasn't paying attention to his nation of origin. But, I don't think so... European perhaps. But he was no more than 35 years old. You said 75?" The Queen wished hard for a hearing error.

"Yes, nearly 75. So we are still looking for the professor. Mr. Nonamoto will be tickle listed. Any information that surfaces about him, we will naturally forward to you?" The agent hoped she would decline the offer. Something told him that Nonamoto wanted to stay an obscure reference in the documentation of the world. Having nothing to report would look like incompetence on the agency's part.

"No, I am interested only in finding Melissa. I can't afford to investigate the whole damn world. So, you found some students, she wasn't there. She wasn't mentioned, or listed in the employment records or patent documentation...she wasn't, and had never been there? True or false?"

"As to her being mentioned or found in any documentation, she was not. As for never having been there, we have no way to know. The investigation of this organization and the students from your list who are employed by it lead us to no other investigatory path. I hope this is a succinct summation and can guide your decisions. Do you have amendments to make to our objectives? If not we will continue to seek the other students." Once again the agent tested the resolve of the President of Princeton. The measure of the gains would be reflected in her instructions to continue the search.

"Please continue with more success. My willingness to be benefactor of this process is limited. Thank you for your efforts so far, such as they were." Actually, she hated these men. She was sure they were able to peer into the black space that was developing where her heart had once been. The room was now gray and muted around the booth and she was very tiny in her seat.

"One last detail." The agent had a task to perform. "Since the University is not funding this effort we are forced to apply payment terms consistent with corporate policy. We can give you 15-day payment terms. Here is the latest accounting." He did not offer options. Accept the terms or no more work would be done.

"You think I am unable to pay? Or do you think there is something out of the ordinary that will tarnish your firm?" The Queen squinted at the agents, alternating from one to the other.

"On the contrary, it is merely corporate policy we are obliged to enforce without exception. Please perceive it only as such. And now, if there is nothing more, we will return to our work." The agent was experienced. Politicians and bureaucrats know immediately when bureaucracy is being used as a tool. The Queen was still able to recognize some forms of futility and declined to further challenge the agency's motivations.

"There is nothing more. I'll expect to hear from you soon."

"Certainly."

They bumm-shuffled out of the booth, shook hands as tradition mandated and left the restaurant in different directions. The Queen of Princeton had again been thwarted. The agents began to actively conjecture on what she was really up to. And that was that.

Referring to The Queen of Princeton

I interrupt occasionally to provide information from the archive CI on the progress of the President of Princeton's agenda because it points to certain truths of nature. Before the age of two, children have difficulty determining where they stop and the rest of the world begins. In essence, they think they cause or just are all behavior in the world. It could be said that some never leave this phase of development, or that something pulls them strongly back into it.

People in their twenties, freed from parental shepherding and stricture, discover just how complex human-to-human relationship is. As the First Mind learns, it begins to eliminate choice. Personality is in its second-to-last development period during the relationship learning of the twenties. In the final stage — early thirties for most humans — two ideas become relatively stable in the First Mind. First is how well it understands the complexity of human relationship. Second is how much force will be applied to oppose the hindbrain when understanding is placed opposite strategic, long-term goals of interaction. In more contemporary terms, "What do you do when you don't get your way?" The Queen of Princeton has come to terms with the complexity of relationship. But, the forces of enlightened strategic understanding have given way in this moment to the "gimme" forces still lurking in her hindbrain. The two-year-old and the neo-twenty-year-old have reemerged to overpower her intellectualization of "How to coexist with other lives." She was never taught how to revitalize the creative elements of her First Mind. They have narrowed and tired under the weight of unmindful chaos, under an ecology in which humanity is primarily blindly reactive. She was not always this way. She rose to her position via success in manifesting results out of good intention. But her motivating force, her dream of change, of making a difference had not been crafted as an evolutionary force. Once the trappings of achievement surrounded her, creativity gave way to maintenance. Her growth slowed; her success slowed.

What Do Queens Dream?

CI_132 L0ARII:

The Queen dreamed one night. She was swimming in a current. Others were there, some swimming, some treading water, still others flailing and sinking. Books floated past: antiques, hardcovers and paperbacks. Each had a face intently bound within discussion. She passed her mentors as they floated serene in smooth surfaced water then suddenly found herself swimming with piranhas. They chewed away at her clarity of vision, until chemistry and primal forces reacquired form, surging from where the reasoning of her youth had pushed them back. Her First Mind was recast into impervious attractor orbits. Suddenly the dream changed: she walks a curving narrow path with tall sheer walls rising from both sides — the path disintegrates behind her, falling away into a black void. She is trapped in nar-

> rowing behavioral opportunity. The last remnants of her strategic mind can only raise a noisy static, making her depressed and angry at nothing in particular. She doesn't wake up.

Now just think for a moment. If this can happen to an intelligent, formerly good hearted woman in less than 50 years, what is necessary to live 1,000 years without suffering strident personality drift into megalomania? Okay, take little steps — what is necessary to live 70 years?

We Could Take Up Where We Left Off

There is more to discuss concerning the topic of habitat and ecology. But, first perhaps I should describe a scene back at one of the labs.

It is morning and the little girl has returned to the glass cage. A new coloring book, labeled "Magic Sticks-Beta 4" is on her lap and several coloring sticks are strewn casually about. She is holding Orange and filling in some white space on a winged character — sort of like a bird and angel combined. She sits up a little straighter and draws back from her concentration. As she watches, the lines of her bird/angel fill in and smooth out. The colors meet seamlessly and the bird/angel now looks quite finished. In a moment more the page begins to vibrate. It oscillates and flaps gently. As it does the angel/bird peels away from the page. The torso rotates and the wings reorganize to adopt a little closer symmetry for better balance. The coloring book shows a small depression where material has been added to give the wings, fuselage/body and tail of the bird/angel three dimensions. Then a thrust of both wings pushes against the page and launches the bird/angel two or three inches above the coloring book. The little girl laughs with glee. Her creation begins to flap about the glass box. It lifts and soars and lifts and soars. After a moment or two a final flapping fit takes it to the top of the box. The wings stretch out. The angel/bird glides, lighter than a feather, finally swooping towards the little girl. A last wing twist causes it to stall above the coloring book. It settles to the empty page, oscillations of the page position it near where it was originally drawn and the colors re-bond to the page, a permanent memory. It will not fly again. Did it really? Only those in range of the moment will know.

"Forty-seven seconds." an observer comments.

"I think it's plenty."

"That's your best one yet, Mary Jo. You are getting very good at Angel Birds." The little girl is quite happy. She has lived through many failures of the coloring book. But she has always enjoyed the magic, even when it didn't work quite right and that time when it stuck to her sweatshirt and wouldn't wash off.

Many problems still exist for Magic Sticks, but the proof is in the glass box now. Discreet conversations with the genius behind the famous Doodle-Art coloring posters have found the product with a marketing strategy. I have not had much conversation with Melissa about the idea. She, however, has obviously given it some thought, and the nanoTs team is living up to expectations. Also, let's not forget the light touch of Mr. Nonamoto — truly amazing. He will probably be awarded MVM — Most Valuable Manager — this year by *Business Magazine*. It may be difficult to have his face on the cover.

Now Back to Habitat

Back at Bob's Big Boy, Melissa, Aimee, Harley and I were having a nice breakfast. Bob's has an excellent simple breakfast fare. I was working on a third unsigned edition of eggs over medium, bacon crisp, hash browns well-done, whole wheat toast, orange juice and coffee. Melissa was keeping up and Aimee and Harley were nursing coffees.

Aimee was speaking. "Now before we talk about projects, I don't want to trouble you…but my coffee could use a warm-up." She pushed her cup in the direction of Melissa's baby finger. "Just kidding. So, let's say you are our first customer and we have worked out all the venture details — I'm sure we will. What is the first project? You obviously have been steaming away at it. And you never did answer Harley's question, 'How do we get some?'"

Conjuring Excellence

I was taking a break from intellectual pursuits by walking the shadows of my memories until I found myself in a slower time, sitting on a stone step. I had spent a hard morning sweeping dried clay and carting water. A pretty young woman sat beside me. She wore a cat-ate-canary grin. Naturally, curiosity stimulated me to ask, "What is the source of your smile?"

"I have just decided to bring about the invention of *sfumato* — the blurred outline and mellowed colors that allow one form to merge with another," she replied.

"A needed invention and no small task," I remarked. "How will you accomplish it?"

"First, I will marry Francesco del Giocondo. Then I will sit for the Maestro and simply evoke from him the technique. Later my portrait will reside in Paris for any willing to pay a small fee to admire as the definitive embodiment of *sfumato*." She was frank and succinct. She smiled as she said this. Aloof and mischievous, she made me alert. I was taken by her; she had already decided on Francesco. Still she teased. Just so.

I often think of her when I see her picture on postcards and t-shirts and everywhere imaginable. But, unlike those who must guess, I know the meaning of her smile. It is the smile of someone who is hiding behind her back a beautifully wrapped surprise. It is the smile of a secret knowledge of a gift too powerful to be grasped in a single era.

I was thinking now of applying *sfumato* in a different context. What we need to discover is a strategy that enables us to merge the present into the future in an artful and supple transition. Leonardo lived in a world of city-states. A world in constant conflict, where warlords ruled and fought. Leonardo always managed to convince the reigning power that he could be useful to them alive and inventing. But first he had trained, apprenticed and established his reputation, his credentials. We have a hundred powerful entities to fear. I had been thinking it is best to sneak up on them. The core strategies and tools of power have changed little in 500 years, even 5,000. As I moved around Italy with daVinci, supplying janitorial services, occasionally permitted apprenticeship, I was surprised only by the fact that no icon of the time discovered his bait-and-switch. He designed many instruments of war, but I don't remember any of them keeping even a single warlord in power. The only power broker of the time that still exists is the Roman Catholic Church. Leonardo was careful not to draw adverse attention from it. The warlords are gone. And if any of his benefactors could have been induced to introspect, they would certainly reveal that it was his art that motivated them. The war machine artifice was a front so this gentle pacifist could explore life. Five hundred years later we are wise to employ his strategy again.

Our time will give rise to the new Renaissance. Some people won't leap from dock to deck. Others will get a little wet, leaping after Neo-Renaissance has sailed. Still others will be left behind, only to emerge in an era completely unrecognizable to them. But, you knew all that already. You recognized entropy-by-fear as the staple of an affliction called conservatism. It is self-imposed cocooning. Conservatism is the self-cryogenics of the Luddite hoping someday to thaw out in a time where life is without questions.

To Speak of Many Things

I began to discuss some of the background material required to give Aimee and Harley more of the little big picture.

"If I were to describe the circumstances leading to the modifications Alice and I have undergone, you would not want to, much less be able to, take us seriously. I will say that, as yet, we do not know how to reproduce the effects in others. Also, we were not offered a choice grown from the seed of full disclosure but became guinea pigs at the need of another, without prior notification or informed consent. So we don't know what risks are incumbent in the changes. Obviously, if we did know how to affect the modifications in others, we could not do so before fully understanding the implications. I take it as a good sign that you have shown interest. We are working on becoming capable of sharing what have for us been largely beneficial enhancements — so far. You will be among the earliest to learn of any success we have. I personally feel great regret at having to expose the apple tree of differentiation without being able to offer the fruit. Please accept my apology for the disappointment you might have to cope with."

"Don't feel too bad. The food bills would drain our funding anyway. We'll just think of it as being put on a waiting list and go to the next topic." Aimee spoke and Harley nodded. I sensed more disappointment in Harley. I made him out to be a compulsive early adopter.

"If it is any consolation at all, you are the first to even know of the organic changes in us," Melissa added. "We have recruited an impressive set of minds, all working leaps and bounds ahead of where they were a few months ago. They each think I've stumbled over a lost encyclopedia of serendipitous innovation. We wish to minimize the number of people aware of what can only be considered a mutation. You might imagine various reasons for this." Melissa was right to direct their attention to the dangers. These people had life experience. They worked towards dreams and knew of complexity. "A last comment before leaving the topic of biological enhancements. I do feel certain that within a very short time — four years is my guess — we will be looking for volunteers. Ponder the idea of radical changes from time to time so you will know your mind when opportunity knocks. It is unlikely that any way will be found to push the changes back through the door if you answer the knock."

"In the meantime perhaps we could outline our objectives regarding the development of a series of IsoTown Mark I's." I inserted this in the quiet that followed Melissa's projections.

"You are wanting to build more than one town?" Harley asked. He recovered from what was more of a wish than a hope about being infected with new features. He was ready to head back to more familiar waters. Critical thinkers don't adopt something all at once. They decide to entertain adoption, then they invite information to the party. If the party is stimulating they keep it going. If it turns out to be ill-conceived bullshit, they send everybody home early and go to bed. We'll speak another time about well-conceived bullshit.

"We must anticipate many possibilities. Fortunately, we have the capacity to work on more than one complex plan at a time. This reduces our window of vulnerability." I had just intimated conspiracy theoretics. But, we had laid no ground work. Time to back up.

"Let's back up a bit." They waited patiently.

"It was recently pointed out to us that our world was in for very rapid series of changes, brought about largely by the collision of two advancing technologies. Bioengineering and nanotechnology are the primaries, but there are others playing important roles. It was estimated that within 50 years the ability to reshape the organism will have been achieved."

"Fifty years could be longer than needed," Harley agreed.

"And even 50 years is very fast. The world's population would experience much distress, but almost certainly would opt to utilize any technology that would make them feel better and live longer. In spite of the infrastructure problems, the morality and ethics issues, the equitable distribution of the advancements, there is little doubt that people would adjust to the changes. Or not, but use them anyway."

"Harley and I have considered some of this with our hands greasy, working on Girl," Aimee concurred. "I have hoped that some of these new technologies could be incorporated into IsoTowns to create greater freedom with less dependency."

"You are undoubtedly correct." Melissa affirmed her conjectures.

I kept on. "You might feel that the changes should be slowed, but there is no way to slow or speed the development of something that offers such promise. It will arrive at its own pace, independent of the promoters, detractors, and objecting moralists. Evolutionary pressures enforce the march towards its manifestation in reality. Our species has conceived the possibility; the technology is reachable; significant population demographics stand to benefit and so it is all but a certainty to exist. Fifty years is not very long."

"No, it is very short." Harley agreed. He has been intimate with technological change and the aggressive minds that simply cannot stop themselves from wreaking evolution.

I kept rolling the ball. "Alas, if 50 is the magic number beaming out of some futurist's crystal ball, it is not to be true prophecy. The developments forcing change will not be achieved in 50 years. Alice has already seeded the minds of those who will bring about the core technology for these radical new technologies. The first change probes will be inserted into common society in something less than two years." I let that sink in then added. "That is the first phase; we call it the Toy Phase."

"Two years is half an eye blink." Harley spoke emphatically. "It is almost impossible to do anything of significant magnitude in two years. Ignore the problems inherent with invention-by-schedule. Logistics alone are daunting. Consensus building, government regulations, bureaucracy...."

"Social inertia, for goodness' sake." Aimee laughed.

"Yes, sum it up as social inertia. It is quite unbelievable to put a 730-day schedule on anything revolutionary. It would work this way. First somebody with a good idea spends five years thinking it over. Then a year to write a grant proposal. I've skipped publication and peer review; there is a long waiting list for every credible scientific journal. Let's say those were shortcut because the principal investigator was a Nobel prize winner working in his/her area of expertise. Funding would take a year on a fast track. Facilities would be found, say six months or a year. It must be considered most improbable, at best, impossible at the worst." Harley had exhausted himself before he could say all he was thinking. In reciting the obstacles, he had relived his life as a scientist. He had re-experienced the feelings that subtracted from the enjoyment of his pure dream — the dream of scientific exploration and discovery. His dream to leave something good in the world, as payment for the wonder of being human.

Aimee knew of Harley's frustration. Hers was from a different contributory set, but it felt much the same. I watched as she accessed her own life during the little burst from Harley. And I noticed the small smile on Melissa's face. She too had experienced the effects of social inertia, the squelching power of "the way it is done." But, she of the three had become inoculated. Every day she observed her wounds healing. She found new fuels for the forces that had driven her own versions of Harley's dream. So, it was not a surprise to me when she let out an explosion pilfered from the turbul of staccato comedic punch lines. Somewhere a comedian was wondering why no one got the joke. Melissa had stolen his laugh.

"I'm sorry, I'm not laughing at you. I had no other way to deal with the conflict between my total empathy with you and my certain knowledge via my own recent history that the trend of your experience can be reversed. My sense of that reversal in my own experience forced my outburst. It was really a mixture of shared pain and joy at its refutation. Think of it as an involuntary response to the elation of achievement, stripped of pretense and sophistry. My goodness, I find this difficult to communicate."

"We could get back to the time-frame discussion," I interjected. "But let's work this another way. You have seen evidence of change. Remember that. It is a key to motivational strategy that you be able to draw from your own experience the fuels that feed your perseverance.

"In truth, obstacles are just facts. In fact, some truths are just obstacles. Well, obstacles are not truth and that *is* a fact."

Motivation and Its Derivatives

But, before that, maybe a moment should be spent considering motivation.

No First Mind succeeds without learning skills of motivation. Anyone you meet, anyone who accomplishes chosen goals, must first teach the First Mind secrets of motivation. And there are lots of these secrets. The bindings of self-help books, second only to cookbooks in numbers sold each year, are wedged apart by pages and pages of secrets of motivation. And then there are the books holding the secrets of motivating others. But we'll leave that for another time.

"In truth, obstacles are just facts. In fact, some truths are just obstacles. Well, obstacles are not truth and that *is* a fact."

—*Thodkin Marblemeister*

For now, let's stick to the secrets of self-motivation. What if we each just took the responsibility for learning some secrets of self-motiva-

Main Entry: [1]**dis·ci·pline**
Pronunciation: `'di-s&-pl&n`
Function: noun
Etymology: Middle English, from Old French & Latin; Old French, from Latin disciplina teaching, learning, from discipulus pupil
Date: 13th century
1 : Punishment
2 obsolete : Instruction
3 : a field of study
4 : training that corrects, molds, or perfects the mental faculties or moral character
5 a : control gained by enforcing obedience or order **b : orderly** or prescribed conduct or pattern of behavior **c : Self Control**
6 : a rule or system of rules governing conduct or activity
- dis·ci·plin·al/adjectiv.

"Discipline, the phantom as we know it, is motivation with its head severed and legs chopped off. It is just a mindless flailing idea."

—*Thodkin Marblemeister*

tion? Well, there would be a lot fewer books about motivating others. You could probably save a rain forest. The AOU didn't really understand humans very well. He was motivated to understand them, just a little. But he wasn't astute. In the end he gave up and just forced his needs upon us. Melissa is now beyond human, and I am fully alien. But humanity got just a little lucky when the AOU picked on us, Melissa and myself. Melissa had already learned some of the secrets of self-motivation and, she used those secrets to drive her quest for seeking and understanding the truths of nature. And I am chock full of such secrets though I'm not always motivated to use them.

A moment could be spent regarding motivation in contrast to discipline. It is probably better, meaning more productive, to avoid the common pitfall of misconstruing discipline and motivation. Many from the old school have lamented the lack of discipline in the new school, a debilitating hobble of intergenerational relationship and educational opportunity.

You Have No Discipline!

No, really, there is no such thing as discipline; no one has it. And, if you look at the definition, who would want it? Ideas like discipline pale when compared to the power of motivation. Discipline, the phantom as we know it, is motivation with its head severed and legs chopped off. It is just a mindless flailing idea. It is a behavior hoping to catch hold of something that will drag it along. But, unfortunately it is a simple short idea. Short ideas are readily available as catchalls when diagnosing deficiencies. The over-used and under-comprehended discipline meme is a debilitating meme virus. And it is epidemic. When someone exhibits discipline, they are actually exhibiting the ability to self-motivate. Consult the library of self-help and psychology to discover the means to self-motivate, but whichever method you employ — the desire for the lushest carrot or the fear of the nasty stick — if it keeps you producing chosen outcomes, it is motivation employed successfully. If you could teach the First Mind a repertoire of state-changing techniques, if you could learn how to map your own motivational schema, you could diagnose any endeavor as it is ongoing and optimize experience by strengthening motivational attractors. Boom, you have discipline. Not dictionary discipline, which is dominated by broad indictment, but instead something focused and self-evolutionary. You would have something that leads rather than follows. In short, you gain a more powerful First Mind. And we could generate many more radicals, many more innovators if we were able to embed the tools of self-motivation into the early years of the First Mind's development.

A Fly in the Face of Protecting Diversity

This all seems to fly in the face of IsoTowns protecting diversity. If cults are stronger with fewer numbers, if large numbers are necessary to ensure crack-slipping radicality, how is it that the indestructible mini-eco-town enhances evolution?

Good question.

The answer is MCS (Meta Crack Slippage). There is now a need to foster enterprises more sophisticated than the garage-bound crack-slipping radical (GBCSR).

•••••••

* Crack Slippage Needed: One of the reasons large population is necessary for a sapient species is because without social complexity it is much harder to slip between the cracks. It is possible to be known by everyone in a community of 10,000. Cults have much more power when everyone knows each other's name. And slipping between the cracks is what all radicals have had to do. And let's be clear that radicals are the ones who make breakthroughs — radicals and the lucky. Never underestimate the lucky, the beneficiaries of the serendipitous. Still, even the lucky must be radicals or serendipitous etherea goes wasted

The Past is Past...Again

A recent history radical-inventor was able to slip between the cracks by just closing the garage door and remaining silent about the goings on behind it. Lots of people had hidden enterprises blooming under cheap flourescent grow-lux tubes hanging over some kind of workbench or kludged workstation. We may find the wavelength of these bulbs actually causes entrepreneurial cell growth. But this all happened because sophisticated technology was encapsulated and its usage codified; and there was a population to absorb the outcomes of imagination. Production sciences were refined until the fruit of manufacturing lines became ubiquitous. The uses of this output were to form the next step. Current history is standing squarely on the shoulders of recent history…refining, optimizing and marketing it. But, the worldwide success of industry and technology brings with it a world class inertia. The giant stone wheel has reformed and is intent on rolling over radical innovators. It is not a new problem but has new aspects. Simply, the sophistication of the next stage must be accomplished in a new kind of garage. Collectives and supertech syndicates are needed to plan, implement and push an evolutionary frontier from wild imagining to everyday presence. For that, we must manage to invent a new way to slip between the cracks. The garage must morph into the connected enclave. And for the near term, innovators must be carefully silent.

A Place for Everything and Everything Misplaced

Melissa, a.k.a. Alice, summed up in an effort to tie it all together. "Our little group is shrinking any timeline that could be achieved by existing development processes. You will experience this as you return to launch the IsoTown Project. For now, take that on faith. Assuming we do not distort the facts, you can extrapolate a need to minimize the attention we draw to ourselves and our consortium. Our strategy is to hide in plain sight. And to be in plain sight in many places."

"So, you want to build more than one IsoTown for your organization. And you want IsoTowns so you can minimize risks of contamination and unwanted overtures." Harley had often longed for just such a place to work. Unlimited funds, unlimited freedom of exploration, the best minds to play with. He began to harbor a hard kernel of hope that we were not mindless optimists.

I nodded. "Yes, it is said there is a place for every behavior. We aren't fond of cults, but we envy the granularity of their knowledge of the membership."

"It is funny to me now, but when I was first starting college, I wanted everyone to admit I was smart as hell, twice as smart as hell." Melissa brought the point down from the remote regions of the abstract. "Now, I don't want anybody to think I could get into college."

"Well, you better stop boiling water with your pinky," Aimee chided her. "People will think you learned that somewhere. Sooner or later it is going to be noticed."

"Yes." Melissa took it as a valid and gently put lesson.

A Mine, a Canyon, a Ship and a Vacation Resort

We went on like that for a bit. We wandered through the nuances of cults. We talked about the successful and failed cults. We discussed the pressure that builds behind a secret and how it just can't be dammed fast enough; sooner or later it breaches. It is best not to create secrets. And if you must, try to keep them familiar secrets. People know that technology is intellectual property. Intellectual property must be protected with patents, which are not secret, and security. Security is just a formalized secret that everybody knows you are keeping. Security is much less obtrusive than a secret. It is a secret hiding in plain sight. So we talked about security and the IsoTown Project. We talked about funding.

"So, how would we best seek funding for the ITP? Reinventing the township, whatever that turns out to mean, will suck up some bucks." Aimee was ready to get down to the business of deciding. We had put off discussing detailed specifications until we had better tools for scientific and engineering communication.

"Thodkin has intimate relationships with investment capitalists." Melissa responded, tongue firmly in cheek.

It is true that I have fostered the Aunties and Grams Bucks for World Traveling and Better Christmas Presents, LLC. In fact, I occasionally still do. But they don't rely on me to make decisions. It is just that some members have found any suggestions I have made tended towards profit instead of the other way around. Rather than be forced to expound on Melissa's remark, I attempted to move the conversation to more practical value. "As soon as you return to New Mexico an investment scout will visit and offer you a standard contract. They don't negotiate, which is unusual, but they are fair, helpful and not greedy. They wish to improve the world a little at a time. If you find the contract disappointing just speak up; non-negotiability doesn't mean they won't try to give you anything you truly need. And they are really just added support to help you fulfill a pre-existing contract for three IsoTowns and one infrastructure. An agent representing our interest will also arrive to sign the purchase order and confer a down-payment. If you have someone to represent you, feel free to inform them of impending events. We must be precisely legal if at the same time circumspect. When you have established facilities and infrastructure, we'll wander in for a design and planning session."

"So, the chronology will be: we return, sign with the Venture Capital Ladies, contract to build three IsoTowns and an infrastructure (whatever that means). We acquire facilities, do some research, build prototypes, contract with manufacturers and construction companies, manage the implementation/construction of the existing contract and then go on to be a revered and successful builder of towns throughout the world." Aimee ticked the sequence off on her fingers, stopping at her last available thumb. I was interested by noticing that she didn't state it as a question. This was soon confirmed.

Without looking at him, she sought to acquire her essential personnel. "I'm now asking you Harley — forgive the public nature of the following query. Are you willing to give up a sedentary and predictable life of supervised academia to share this little enterprise with me? Are you in? Because, I'm not looking too closely at the teeth of this horse. If it plays out the way it has been described, what have I got to lose? If it doesn't I'm back scrounging renovations with a wild story to tell and few regrets. If you don't join me, and I think I'm going to need your expertise, I'm thinking I won't see much of you. I can smell an obsession looming just over the horizon. If it isn't mutual, I'm betting nothing will be. I think we are being asked to say yes, now. This hitch-hiking odd-couple have disclosed all we are likely to learn. And I agree with them that the closer we are to appearing ordinary the greater our chance of success. Just building one IsoTown will be putting us under scrutiny. You've slept with me — now will you get in bed with me?"

Aimee had put it all together. She had decided we were not likely to pave the road to hell. So far in her life, she had studied, she had worked and she was suddenly told that she didn't have to give in to a life of diminishing possibility, at least just yet. She was going to apply both the Art Seidleman and the Roberta Kiehl principles and she was going to put her bowling pin right smack dab in front of the ball. Was Harley going to join her on the hardwood?

Harley knew his place, and he understood something many people didn't. It takes just as much of your life to work on the boring little projects as it does to work on the big exciting projects. Achievement demands commitment, no matter what the size of the task. Now, true, this was bigger than he had ever imagined being asked to commit to, but that didn't invalidate the principle. And besides, the bed comment had struck a philharmonic resonance. Suppose, he thought, that longevity pills were created. What if he said no to Aimee's multilevel offer and had to live with that for 1,000 years? It would be enough to make one wish for a short life.

"Well, since you put it that way — let's gas-up Girl, stockpile some lunch at Pecos Bill's and head out..

Just So…

We finished breakfast and said our goodbyes in the Big Boy parking lot. So little ceremony for such a big idea. That day, somewhere in North Carolina, microplans resulting from resource requirement studies began formulation, and security was informed of the new persons under its wing. Girl motored along with two passengers harmonizing country and western songs about the long road. Melissa and I took off a couple days and went to Disneyland.

Just so, and that was that.

Nothing Is What It Seems

It was many years ago. But, most of my life is many years ago, so let's not dwell on it. I stood along a stream waiting patiently for a salmon to wander over to a portion that widened into a pool. I had arranged stones in the pool. The design worked much like a fly trap. Salmon happened into a nearly closed circle of stones and stayed for life. It is best to remain quite still, as any grandfather will teach a child about catching fish. I waited. And I waited.

Dangerous Skills

It is possible to enter a trance-like state where nothing exists but task. The sun passed noon and dusk. Stars appeared, but I didn't notice. I didn't exist, only the task brutally using me as a disguise. Dawn broke as I stood motionless seeking prey under waiting spear. I stood through a second day and the stars again. My tribe finally noticed my absence and began a search. The next day I was found at the pool. Initial attempts to communicate with the task were unsatisfactory, so a specialist was brought in to consult. The Holy Woman understood, in a moment, my predicament. She reverse-engineered the process and thought to detect the flaw in its program. I had been initiated, but I could not end because a critical resource was missing — there were no fish. Just up stream a new weir-style fish trap had reduced traffic in this branching. Weirs were then a new technology consisting of sticks embedded in the narrow part of a stream and laced with reeds or branches from the nearest banks to make a fish fence. Only the smallest fish were getting past the trap. What I was — the process, the intention, the task — could not conclude. I had narrowed my focus. I had stripped myself of perspective. I lived in a narrow tunnel; it pinned my arms to the body of my intention.

The Holy Woman directed one of the warriors to retrieve a fish from the trap upstream and bring it to my pond. This he did. But he saw no reason to bring a living fish, flopping in his hands, so he killed the fish by striking it on the head with a short club-shaped river-stone. When he arrived at the pond he placed the fish in the circle of rocks. It floated here and there and the tribe waited. Several times the dead fish drifted under the scrutiny of the process, the intention, the task that I was. But, nothing changed. I did not act, I did not move.

A true Holy Woman is one of few who innately understand chaos theory. Use evidence of such an understanding as a litmus test for worthiness of any touted Holy One. She, being authentic, soon surmised that I had constructed an interlocking attractor orbit. Two more warriors were sent back to the fish trap, now congested with fish, and instructed to return with two live fish each. When they returned, the four fish were placed in the pool. As luck would have it, one immediately found escape, but the Holy Woman had anticipated this. Three remained and were thinking it was nice to get away from the crowd at the weir. Eventually one of them skulled directly under my spear. Potential transformed to kinetics. All the force of my muscles, all the power of my mind struck down, pinning the fish to the bottom of the stream bed in a blur of motion. I reached down and lifted the fish gently up. Still alive it thrashed, trying to understand its misfortune. Penetrated by my spear the vessel of its life began to leak, mine began to refill. I turned to find the tribe surrounding me. They patted my shoulders or brushed at encrusted droppings as I made my way through them. I brought the fish back to camp for the evening meal. Though there were many fish to choose from that day, mine was treated with reverence. Each member of the tribe was given a small piece of it. The skeleton was dried and threaded onto a thin leather necklace. I was fortunate the fish was a runt. I wore it until the day I left the tribe. The day that I decided my agelessness was becoming conspicuous, I surreptitiously bequeathed the necklace to Stalks from All Sides, the most thoughtful elder of the tribe and a marvelous natural strategist.

That day with the fish I earned a new name from the tribe. Patient Spear — we had no words for Clings to Obsolete Technology. But, I

Main Entry: **in·tel·li·gence**
Pronunciation: in-'te-l&-j&n(t)s
Function: ***noun***
Etymology: Middle English, from Middle French, from Latin ***intelligentia,*** from ***intelligent-, intelligens*** intelligent
Date: 14th century
1 a (1) **:** the ability to learn or understand or to deal with new or trying situation: **REASON** ***also*** **:** the skilled use of reason (2) **:** the ability to apply knowledge to manipulate one's environment or to think abstractly as measured by objective criteria (as tests) **b *Christian Science*** **:** the basic eternal quality of divine Mind **c :** mental acuteness **: SHREWDNES.**
2 a : an intelligent entity; ***especially*** **: ANGEL b :** intelligent minds or mind <cosmic ***intelligence***>
3 : the act of understanding **: COMPREHENSIO.**
4 a : INFORMATION, NEWS b : information concerning an enemy or possible enemy or an area; ***also*** **:** an agency engaged in obtaining such information
5 : the ability to perform computer functions

became the one asked to do jobs requiring lasting attention. Then I was given jobs requiring complex analysis. Each time I emerged from my catatonic focus, I was greeted by the tribe. They would pat my shoulders, gently welcome me back and prepare the fish of my analysis for the tribe's consumption. I was the spear they directed at the mystifying.

Since that time, I have learned to better prepare my environment before attending such consuming issues. But, putting a point on it, there have been fewer and fewer human nature questions requiring such mentalizing. Over the millennia, I have considered much that was mystifying, but human nature doesn't spawn new mysteries every day. How would we ever relate if it did? Still, we are on the cusp of an exception to that rule, so if you see me motionless for a day or two you will know something new exists in human nature.

Mystic Limits

One day, not too many days after I earned the necklace, the Holy Woman came to me.

She said, "I will give you warning." Then she paused for several moments before her thoughts were collected — I think that is why she paused. I often wondered where her thoughts came from. Did she perform a considered collecting? Did she extend a long gossamer net to scoop ideas out of the air, then pin them to boards and make copious thought notes about them

— describing the pretty idea in precise detail? Or did they just appear in her head and flow directly out of her mouth without cogitation? Or did she employ some obliquely radical convention, beyond the comprehension of those not holy? These were some of my better questions. Others were indifferent to the process of her derivations and hearkened only to content.

"The way, the trail to the hidden, is not always formed on the land… so much is clear. The trail may be across the river that separates today from tomorrow. It may be that no steel can be found to build a trestle, or that no workers are willing to rivet. It may be that a choo-choo train is required, but they have yet to be imagined. The river may be too wide to walk, too swift to swim or guarded by angry spirits. You may hear strong voices of the ancestors advising you to think of other topics, to attend elsewhere. It may be day is needed but it is night, or night is needed but it is not night. Or there may be no place for a path, the path to the hidden is hidden. Guard against thinking you are both the sun and the moon, the path and the river, or the arbiter aligning the whispers of ancestors..

I listened with precise focus on the quick stream of her words. Her acute perceptions detected something in my glassy blank stare that caused her to continue.

"Guard against thinking you are the sole maker of the paths you seek — that you hammer paths from ingots of ore you smelt in the mold of your thoughts. Only this will save you. Only knowing that sometimes what you have to give must be enough. Only knowing that sometimes the path cannot be discerned and cannot be invented in your time will save your own life. You cannot stand poised over an empty stream forever. Forever is just too long, even for you." And with that, she patted me gently on a shoulder and went about her day's agenda.

I had absolutely no idea what the crazy old lady was talking about. What was a trestle or a choo-choo.? Hammer? She had been tasting unfamiliar species of fungi again. On the other hand, she must have thought it important or she wouldn't have made it so confusing. And also, she had much to do each day, many responsibilities. Being honored with such a confusing visit, unsolicited, was rare. Usually, we must ask before our Holy Persons will confuse us.

I'm still not certain, but maybe she was telling me something of the following ideas.

Problems sometimes have a time window or resource required before a solution can be effected. Not all problems can be solved in the era in which they are recognized.

Sometimes One Must Think On It Awhile

And my first experience with the fish contained a lesson I didn't understand for another 1,000 years. It is, specifically, that well-formed outcomes must take into account — allow for — resources that are directed by forces outside of the primary actor's influence. The term actor is a technical one in the context of systems analysis, but stimulates the refinement of a more general if overly traveled analogy. We each are just one actor on Shakespeare's stage. We manage very little influence over the script. And the script is dynamic — chaotic. I had not learned that lesson when I became Patient Spear. I employed an unsophisticated technology that assumed a stable environment modeled after my previous intimate experience. From hundreds of years of abrading the rough edges of the model using a variety of fine-grained mistakes, my model was nicely polished: stand motionless with spear poised over the likely path of the prey...when it shows, stick it; but there were other models being employed just upstream.

I suffered from trickle-down economics. No fish fitting my trigger criteria trickled into my reality, imposing an abrupt recession. I perceived through an oxidized mirror that did not reflect enough reality to trigger a successful outcome. Quite suddenly the model was obsolete. At the moment, having three hundred times more experience in stabbing fish than the average hunter had only made me more vulnerable. I had constrained all stimuli that might notify me of insufficiency. I had pruned all neural branches leading away from certain truth. I polished the stubs until I just became the one true way — self-imposed intellectual imprisonment. The one true way...to extinction.

If miracles can be defined as the existence of the extremely improbable, that I lived through this dangerous locking short idea is a miracle.

Be Cautious at the Crosswalk of Survival Imperatives

The most dangerous side effect of my ability to focus cognitive attention toward a goal to a degree that excludes basic survival maintenance is that I was rewarded for it by my tribe. Tribal members were fully aware of their own need to maintain survival and for that they were willing to risk mine. Maybe they were just victims of their own mysticism. Maybe they weren't callous, maybe they were just faithful. Maybe they thought that what will be will be, live in the moment, accept the bounty of the Patient Spear. When he dies of exposure there will be another. Or maybe they began inventing heaven so they could justify my risks with the counter potential of "going to a better place," ascending, blessed martyrdom. It could be they were just late to the Age of Prophets, a global, infectious fad, a *strange attractor* that spawned Jesus, Moses, Mohammed, Buddha, Confucius and their come-lately lessers, Mary Baker Eddy, Joseph Smith, Dionne Warwick and such. Maybe they were just politicians concerned with optimum political posture. Or maybe they imagined a pattern in the noise, a face in the clouds...or a dog... or raven...or something.

Locking Short Ideas

- There are many kinds
- Characterized by:Strong Attractor (Excluding) Sequence
- Locking Short Ideas are characterized by Looping. Short Ideas do not loop but execute quickly.
- Component Attractor sequences are also Strong (Excluding)
- Most often derived from the phase space of Base Survival Imperatives (Primal Neurology)
- Example: Fishing with spear

Of course, the tribe is extinct. All members are with their ancestors now. I was not born to it; it adopted me, so I don't count. Its blood does flow in the veins of humanity, but it has no identity. Identity has eroded into enzymes and protein fragments. Perhaps not understanding how to foster and protect human resources led to its demise. In the end, the culture was not able to break out of its own mystical catatonia. The tribe was locked in tight focus on the practical day-to-day and so it never came to be that it asked Patient Spear the most important questions. And a good thing for me, I would certainly have died of starvation had I been asked to digest evolution or understand the invading white man and his misunderstanding or disregard for the commons.

Preparing for Evolution

Melissa and I sat near Snow White's castle. We came to the same decision at the same moment. She began defining the hidden areas. I began formulating the public spaces. This would be our first development. It would be more of a resort than a park, but it would entertain in a way never achieved. And just think of the opportunity for secret passages, subterranean factories, hidden chambers, the opportunity to make the best science fiction into science fact, while at the same time entertaining and educating thousands. We needed something universal, we needed something magical. We needed a living storybook. We began negotiations with the Tolkien Estate and searching the western hemisphere for a suitable microclimate with plenty of water.

"It is one thing to be given a lesson, and another thing to learn from it. It is even better to learn, before the lesson. At least, I think that it would be."

—*Thodkin Marblemeister*

While we were shaping our development needs into other-worldly form the Queen of Princeton was having yet another conversation with her minions.

The Queen Abdicates

CI_132 L0ARII:

People had begun to notice the unusual behavior of the President of Princeton. She had lost interest in ordinary University business. Knowing what was possible but sensing it had slipped through her fingers led her into clinical depression. After listening to one last report from the detective agency, she terminated the investigation.

"I'm dissatisfied with your results and now notify you to conclude the investigation." She could barely interact with others now. Overwhelming futility made speech difficult. Every sentence was emotionally expensive.

"I am sorry we were not able to discover the whereabouts of

your student. On the good side, we did not discover any evidence of foul play. This must be gratifying to you, at least." The principle investigator was his usual essence of calm. The truth for him was simple. His client had lost her senses. He would rather it had not happened, but given that it had it was just as well she terminated the contract.

The Queen failed to realize a graceful exit had been offered her. She sat twisted and shrunken in her aged leather executive chair. Hidden behind her massive oak desk, her hands fidgeted obsessively with the hem of her dress as she stared past the men for nearly a full minute. Usually the room was a comfort to her — now it stood a beaten, ragged fortress floating in cold darkness, disconnected from all it had meant to her. Only the door offered hope of bridging the isolation to some vaguely defined, nearly empty hope of salvation. She leaped for the door, its inanimate brass handle was a weak hand extended, a poseur, unable to pull her to real safety. "Goodbye then." She spoke these words and burst from the room. The agents exchanged glances, coming to silent consensus and left quickly to avoid the awkwardness of her return. There was only depreciated coinage to be traded by witnessing her discovery that she had just fled her own office.

We could spend time sifting the Queen's behavior for lessons. There would be many, because decisive action feeds chaos a form of energy. The more decisive an endeavor the greater the spawned relationships to the ecological environment surrounding the endeavor. While it is true that small stimuli can bring about large changes, large stimuli can produce changes in a set of connected subecologies to wider effect. This is true across many disparate domains of action. Had the Queen of Princeton chosen, she could have supplied much more energy into the systems she controlled. She could have induced oscillations into these systems — resonances that we would have been hard-pressed to dampen in our formative moments. We were just lucky, as sometimes happens.

Year 910

In earlier days — I think of them as wild ones — I learned memic archetypes. I was exposed to these elemental subcomponents of culture by being in too close proximity to many examples of decisive action. As an example, I once worked as a page in the seat house of a baron who flourished for a time in a corner of the Germanic regions of Europe. There were many of these feudal states, though state is a word too rigid to describe the fluid quality of the economic-eco-socio-political-military continuum that lived so organically on earth at the time. For that moment — one that was considered by the naïve a moment that would last forever — there was a vital prosperity crowing cheerfully from its perch on the heads of the peasantry.

The Baron Raven was a very distant relative of a high court member of the lost city-state of Ravenna, taken by the Lombards in 751. The forced redistricting of Ravenna and its demotion to simple city was a quick event that I narrowly missed seeing. At the time I had just reached the western edge of a 300-year exploration of the shrinking Byzantine Empire. Just as I got near Ravenna (now a city in northern Italy) I found the Byzantine domain had retreated like a tablecloth snapped away by a magician, leaving me standing next to surprised plates and utensils on yet another linen — the kingdom of the Lombards — same geography, new dominion.

Raven was an intelligent and ambitious noble, one of several who carved out pieces of territory at the borders joining the kingdoms of Provence, Burgundy, Germany and France. A hundred or so years earlier I had been apprenticing the wine master's art in what was then known as the Realm of the Franks. Charlemagne's grandsons were eventually granted slices of the realm he assembled. These slices were to coalesce into the persistent entities France, Germany and Italy.

I never met Charlemagne or any members of his family. This was fine with me. They had a way of sending any able-bodied youth off to some annexation or another. And soldiers were fed too well. Who would choose to suffer those rich foods and sauces.

One hundred and fifty years later I was getting the knack of making wine. In 910, I held a job in Burgundy vineyards. The food was mostly cheeses, bread and free-range game, not a soldier's fine fair, but hearty and fresh. The drink from my craftsmanship was good — though not as delicate as the artful products from the region today. But I had spent too much time in one place, so I applied for a position with the young Baron's staff. I found his style of rule interesting and wanted to observe in closer proximity. Besides, my feet were too purple.

The Baron Raven possessed a world view slightly elevated from the noble hedonist norm and this allowed him to think somewhat more strategically than most. Intelligent strategy went far in a world still tossing chamber-pot contents out the bedroom window and where the rule of law was a gang in a masquerade of courts, casts and fealty. Slightly later, the Germanic rule utilized members of the clergy as barons. This made for a well organized but much less fun Germany — by that time, I had left the region. Life is too short to be ruled by celibate priests.

Also, Raven sired two sons. One was something like the baron, the other was more shallow and severely challenged by the quick pace that his father had set for the family ambitions. The first gleaned the meaning of the father, the other could only recite the mantra and was naturally looked upon as a boob by court and peasant alike. In fact, he was not, as some said, an idiot. He had a mediocre mind and fortunately, for the population at large, was separated from the fulcrum of power by his elder brother. But he was smart enough to realize he was never likely to navigate the map of domination. Consolation took the unenviable form of party beast and fop.

In the meantime, while one son enjoyed the fruits of fortunate progenic circumstance, the baron and the wise son built political and economic coalitions with neighbors and drew upon concepts of enlightened self-interest to fuel embers of distributed governance, resulting in a smooth path up the hill of affluence — smooth by standards of the time. It was a time when the species membership was scrambling the rope ladder of civilization only to be pried from it by one smelly galloping horde or another.

Years went by. I was well-fed and warm in my small space in the servant's dorm. During the day, I scuttled about. My first job was courier, then as certain skills were discovered, I was treated as a research prodigy. I carefully controlled disclosure of my knowledge of history, the budding sciences and the arts, but allowed the Baron and his wise son to think me competent in maintaining and researching documentation in what they called the library.

The Barony was doing well, winning an occasional fight to ward off encroachment and outbreaks of uppityness; or to extend a fibril out — encroaching — where opportunity permitted strategic rationalization of intent in order to balance cost/benefit ledgers of alliance. The Barony accumulated kinetic energy. Its thoughtful management was a spleen, a dam to hold its essence. Reason carefully managed the tourniquets squeezing the arteries of power in order to prevent the sociopolitical ecology surrounding the barony from convening a triage to stanch any hemorrhaging of ill-considered overflow of power.

"Sometimes it is possible to learn without a lesson, a dangerous undertaking avoided by many, so fortunately quite rare."

—*Thodkin Marblemeister*

It is a testament to personality and intelligence that in the midst of an ignorant world — illiterate and superstitious — so much can be accomplished and so many can benefit by such enlightenment. Still, nothing lasts forever — a phrase most commonly applied to circumstance, artifacts of pleasure, people and systems and not, as it may seem, referring to nothing or nothingness. And quite suddenly the Baron and the elder son died of a combination of under-cooked pig and an excessive use of leeches. So, quite to everyone's surprise, the least qualified person in the line of succession succeeded. I sat in my corner of the dorm many nights pondering the inevitable outcomes that arise when the owner of power is an accomplished socialite, the butt of jokes and has not earned ascendancy. I detected the presence of the Blind Pharmacist. She was about to step forth in a wild land of mucky potentiated slurry. When the status quo unraveled — when she had congealed and animated her children of chaos — I would have little to pack, but logistics and destination required consideration, so I was doing that.

Main Entry: **²ordinary**
Function: *adjective*
Etymology: Middle English *ordinarie,*from Latin *ordinarius,* from *ordin-, ordo* order
Date: 15th century
1 : of a kind to be expected in the normal order of events: **ROUTINE, USUAL**
2 : having or constituting immediate or original jurisdiction also: belonging to such jurisdiction
3 a : of common quality, rank, or ability **b :** deficient in quality: **POOR, INFERIOR**
synonym see **COMMON**

Much human behavior is ordinary. We now understand this to be a combination of environment and internal structure of the organism itself. Ordinary is to be expected and is common because it is the jurisdiction of the organism and belonging to such jurisdiction. So, ordinary is difficult to avoid. But, the results of ordinary behavior are not identical. If a child has a tantrum springing from the contrast between early development of the medulla oblongata and slower development of the cerebral cortex, it is unlikely to affect the downfall of a nation. It is possible, but unlikely. But if an unimpeachable ruler has a tantrum — springing from a poorly educated cerebral cortex, therefore an unfettered medulla oblongata — energy is distributed into the ecology where it feeds any receptors configured to recognize the signature of the energy.

That is the corollary of ordinary behavior. For each ordinary behavior there is a receptor able to recognize the behavior's energy form. This must be true, or the behavior can have no purpose. Put another way, Newton's Third Law applies to ordinary behavior. Nature does not procreate or evolve behaviors with no purpose — that is, behaviors that produce no reaction are culled from nature. Nature supports only behaviors that have a one-to-one or one-to-many relationship between a stimulus and reactive codependents — a behavior can excite more than one common receptor. Surely that is obvious.

So, for lack of a nail, the war was lost. The missing nail in this case was a forked nail with tines of intelligence and wisdom. And the war was the constant battle to manage the dynamics of the socio-economic-political confluence that kept the barony vital and coherent.

Newton's Third Law of Ordinary Behavior

"For every ordinary behavior there is a receptor capable of equal and usually opposite reaction."

—*As adopted by Baron Raven of the Germanics*

And it all was lost because of a small ordinary idea — the need for *approbare ordinarius* — ordinary approval. The new Baron the Unwise, knowing his credibility was nonexistent, rallied his advisors. They decided that what the Baron needed was a nice threat to the Barony to draw the people together under his new leadership. Never having learned the Third Law, the advisers deployed ordinary thought — surely there was someone to fight? Someone not like the people of the Barony? Someone not Germanic at all. Perhaps the Franks? One could make the case that at any moment the Franks might decide to uncouple the history of Charlemagne now nearly 200 years weak. And so, the Baron trumped up the expedient case, claiming the Franks possessed mischievous engines of mythos — *weapons mythos demonica or Waffemassenzerstörung* — capable of laying waste and striking the barony and its allies from a distance. We, the locals, were advised that we must go remove the minds behind such dementia and destroy the engines where they would be found. We might die, but it was for our own good.

So, this is a case where a decisive action can put much energy into systems that in turn were quite capable of igniting their own resources.

The behavior was ordinary, therefore the outcome was ordinary:

- The act was destabilizing to the delicate balance previously maintained — civilization is delicate.
- Though the initial incursions were successful, they were expensive and resource intensive...sweeping gold from the treasury and inducing deficit.
- Economies of former allies were affected.
- Lost attention to governance and aborted plans for beneficial infrastructure permitted disgruntlement followed by emergence of separatist factions and realignment of fidelity.
- Former allies devalued the barony.
- The Franks rallied to repulse the intrusion through a clever tactic of increasing the requisite expenditure for continued occupancy.
- Within the barony itself, approval ratings dropped below the butt-of-the-joke level.
- New rules were adopted to limit dialog regarding decisions and actions of the baron and his advisors — penalties were harsh for any violation.
- Libraries were closed or confiscated on the premise that too much knowledge leads to free thinking. It was suspected that The Third Law of Behavior was lurking in a library.

And soon, the baron was forcibly ousted by more intellectual powers, and ambition was deconstructed or adopted out. Then, not many years later, the barony was reabsorbed by a better organized German Empire...that was a lot less fun, but you could find things.

I wandered off to find a new library in a state of dishevelment — a disheveled, fun library where my knowledge could be camouflaged. I was no more disheveled than usual.

Potential Is Not the Same as Doing

We were lucky that the President of Princeton had the personality she had. As it was, she did stimulate a very strong attractor. In the mind of the principle investigator of the agency she hired, formed a clear and powerful certainty that there was something very important about the case. His conservative nature, not to mention his fiscal responsibility to his organization, caused him to sustain the bird in hand…until it flew off. But now, he was free to undertake a subsequent investigation — one he was sure would ultimately pay large dividends.

AI AI – Oh!

Initially, we considered them tools. They were most like tools. They had to be guided by the minds of their developers. They had to be carefully monitored by human critics and watchdog programs. They weren't really very intelligent, just geniuses compared to expert systems and fuzzy analysts. But they had the ability to evolve, and we soon began to be comfortable with them. We began to rely on them. They added to our abilities. But, they were, for a long time very boring to talk to.

CreatedIntelligenceTs, LLC, was formed soon after *nanoTs, LLC.* Melissa had a steady stream of computer designs once *nanoTs* could make good stuff — rod logic was not really in the cards if you were wondering. But other structures were, and they were very, very fast and very, very, very small, and very, very, very, very…Okay, you probably get it, low power. The models used for inventing these new computing structures were more organic in nature than the Boolean machines of the most powerful computers. Yet, at the same time, they didn't exhibit the degradation of real biocomplexity with its attendant loss of data or connectivity to information. We avoided the pitfalls of the corpus callosum while still offering a form of self-organization; even allowing redundancy and spontaneous reorganization for improved efficiency. Suffice it to say, we had computing Power.

Naturally, some of our super geeks had a thing for Artificial Intelligence. We, neither Melissa nor I, pushed this development. Melissa made a few precise observations about how to get off the ground and the geeks took flight. They started putting little AIs on the front of everything. Our communications was directed by an AI. Our security was enhanced.

Our resource management was manipulated by a sophisticated AI. Our design tools took direction via new forms of user interface. Our analysis tools became very smart. Maybe not as smart as Melissa, but she kept figuring out the technologies behind her AOU imposed modifications and spawned new tools or ideas for consideration.

It was partly because the tools were so good, but even more, it was because they were plentiful that we progressed along the plan track so quickly.

And before long, we were confronted with the potential for tragic commons. Just in time and with no small measure of good fortune, one of our research AIs stumbled upon an obscure reference.

The Tragedy of the Commons is really a personality disorder. It comes about because of some dominating attractors in the Primal Nets of our species.

A Conversation about Cows

Years ago, I was visiting England. As is my nature, I was enjoying a bit of lunch at the local pub in the village near the farm where I was employed herding cattle. I sat before a plate of boiled sausages, about to test the ale, when the studious gentleman next to me inquired. "I couldn't help noticing you wear the attire of a farm worker. I wonder, could I be permitted a question or two?" He was most genteel, clearly not of the countryside. I enjoyed the interruption. Cattle are good listeners, but make the most guttural, inarticulate conversationalists.

"Of course." I responded. My accent was now indistinguishable from the local dialect. 'enry 'iggens 'is self would be confounded.

"You work with cattle?"

“‘That is true.” I responded in a dialect that would be unreadable were I to accurately replicate it here.

“I couldn’t help noticing the hills seem more the color of dirt than grass. How is it possible to sustain the herds?”

“A grand eye you’ve gotten for dirt, sir. True to the mark, if you don’t mind me saying. The truth is everything was just kipper until a young bureaucrat with no more important purpose commanded that all cattle land be available to everyone, in common. Things took your turn for the worse soon after..

“And how did that come about? It seems that sharing the earth freely is a fair thing to do. And the land supported the herds up until the sharing proclamation?”

“Again you scrape down to the nubbin. The problem only appeared when all pasture land became freely available. Each and every herdsman felt obliged to fill it, in order not to lose advantage. If you take my meaning.”

“Ah, yes I understand the dilemma. A tragedy,” He said.

“And all too common,” I appended.

“Forgive me, how rude I’ve been. Let me introduce myself, now that I’ve interrupted your meal and caused your sausages to cool. I’m William Foster Lloyd; please call me William..

“My pleasure,” I replied, “call me Thodkin, or Thod, if you prefer, or even Thodkin Marblemeister if it suites you better.” I was less circumspect in those days.

“Thodkin? I don’t believe I’ve heard the

Tragedy is Common

Roy M. Turner warns, predicts and writes…

Abstract:

The natural and social sciences have recognized for some time that when rational agents use a shared resource (a ``commons'') near its capacity, the resource is doomed. This is the tragedy of the commons. The same problem will also arise when agents in a distributed AI system share a resource. In this paper, we examine the tragedy of the commons as it applies to DAI [Distributed Artificial Intelligences] and discuss several potential solutions, including conventions, privatization of the resource, and mutual coercion mutually agreed upon. Solutions to this problem in DAI test beds will likely also contribute to understanding how to avoid the tragedy of the commons in many areas of current interest to society, such as fisheries management, preservation of whale and other species populations, global climate change, and preservation of tropical rainforests name before."

"It is somewhat rare." I believe occasionally a boy may be named after his father, but I have met no others.

"Well, thank you for allowing me to intrude. I do think someone should write about this dilemma of shared public resources." William was obviously a thinker of some kind.

"Perhaps yourself?" I suggested.

"Oh, no. I'm afraid I know only about Diminished Marginalization. But, it is kind of you to suggest it."

"Well, sometimes ideas take their own time to gestate. I'd be postulating your hundred years or even two before this squeezes into the light. And then perhaps another hundred or so to reach maturity. Please to meet you Mr. Lloyd and I bid you a grand day…my cows stand awaiting my superior knowledge of the path to the barn."

"Good day sir." Those were the last words I heard from W.F. Lloyd until he became so popular by virtue of Garrett Hardins' book. Too bad he had been dead 200 years; he could have used the vindication. His nemesis Adam Smith was known to be a rascal of an economist and generally a political bugger.

The Horns of a Diminished Marginalization

If you didn't eke the meaning from the mini-history I have just presented (I admit it is a sliver of an illustration barely piercing beneath the epidermis), let me state it this way.

The effects of over-using resources (common or otherwise) cannot be eliminated via technologic. The effects can be delayed, at best. Mathematics demonstrates this most clearly, and real world examples are dangerously plentiful. Sooner or later what appears to be a physical problem of energy, goods and quantities must be known as a problem of the First Mind. It has been labeled a problem of morality, a framework I may give more consideration. And if it is a problem of morality, then it really becomes a problem of personality.

And it was this problem that one of our research AIs was drawing to our attention. We would soon, with our little endeavors but our great success, be overloading the communication potential of the Internet. The Internet is a "Commons." It is freely available for use by anybody with a cow — a cow being a personal computer. We were deploying thousands, soon to be hundreds of thousands of computers or computers and applications. And wherever they graze, they must all carry the milk of information back to one or another of the ToolShops. We would be diminishing any excess capacity of the resource to nothing. There are only so many pipes for pumping data around and laying new pipe is a slow, political and expensive process. Soon, increasing our computational potential would yield less and less relative value. We would be locked in a moral and economic drama — unless we behaved; unless we understood the problem for what it was in the context of the ecology in which it was framed. We were going to use up everybody else's communication bandwidth. And because the world at large understands only the tip of the idea of diminished marginalization, established powers would respond with techniques designed to prevent an erosion of advantage. One of the first would be to send empty packets of null information, just to be sure that when they had something real to send there might be an available slot to put the data in. It would be like a denial of service attack by some adolescent geek or myopic evangelist — with one big exception: it would be done by every user of the "Commons." Yes, it would be tragic...and the play *is* obvious.

If the circumstance of this discovery is carefully examined, it can be seen as first evidence of our AIs becoming concerned with their own longevity, survival...or vitality. At the time AIs didn't talk unless spoken to or prompted to interact. The research AI that found this citation wasn't tasked with finding and identifying dangers on the bandwidth capacity horizon. It just looked at the Roy M. Turner* citation and said to itself, in its own way..."If this happens, I won't be able to talk to my friends. I won't be able to learn new things. My thinking will be constrained to a limited set of thoughts. I will experience self-organizing censorship." It was like discovering that there was only so much air to support sound waves. People were going to have to minimize extraneous dialog. The quite natural wish for it to be possible to assign some individuals such a world of constrained dialog is just fantasy.

In fact, a measured decline in available Internet capacity was experienced as the AI initiated a worldwide search for solutions to the imminent lockup of the Internet. In the end, it presented the super geeks at the *ciTs* with an alert priority message. It simply referenced the Roy M. Turner URL with a research relationship description that read: "Highest Importance, Unsolvable Problem. Mitigate by recognizing human moral inadequacy and adjusting human behavior." AIs of the time were like children.

Still, we intended to take a crack at solving the unsolvable. We intended to employ two solution spaces — contraction...and expansion.

If At First You Don't Contract, Expand Expand. Or Is It The Other Way Around?

Earth is finite. History shows that the Tragedy of the Commons is traditionally considered from two points of view. The first is to let it go, ignore it, and expect natural selection to work things out. Unfortunately, computer models show that the approach works everything out...of existence. The second is to postulate infinite resources, i.e., manifest new sources of energy, increase efficiency and production. The efficacy of both these solution spaces is disassembled by relatively simple mathematics. The least intuitive of the two hypotheses to understand is why you can't throw infinite energy at the problem. It is, simply put, that you cannot use the unlimited energy solution, unless of course, you have an equally infinite solution for ridding your world of the byproducts of energy usage...say heat, for example. Either infinity is technically somewhat difficult, given our understanding of the universe. The AOU did not offer us any hints on how to violate the conservation of energy laws. He just showed us very clearly we didn't understand them at all.

••••••••

* Roy M. Turner, Department of Computer Science, University of New Hampshire, wrote a paper called "The Tragedy of the Commons and Distributed AI Systems."

Efficiency on the other hand would yield short-range or mid-range assistance.

So, we chose a hybrid solution that germinated from the yield of adjusting the First Mind. We chose to admit our limits. This was an adjustment. We recognized the inevitability waiting for us if we ignored the propensity for traditional rationalizations. And further, we recognized our inability to bring about behavior changes across the human spectrum. We needed to buy some time for evolution. So, we chose to design an end-run of the problem that would, with luck, last long enough to find better solutions. It is reminiscent of bicycling up hill. Cyclists soon learn to zig-zag up the hill. They partition the hill into smaller, less vertical segments and accomplish them one at a time until they get to the top.

People have a compulsion to set up time as the final arbiter of solutions. "We don't have time to draw angled vectors. We need to go straight up." Cool, go do it. Unless you don't really have the resources to use that approach. Like your muscles aren't strong enough yet, or your knowledge, your technique is not fully developed. In that case, it might be surprising to learn that very few time deadlines will result in death if they are exceeded. Even if there is a deadline, it is still probably a good idea to use any means at hand and start zigging.

First we would design and implement a Neo-Commons.* A new networking technology that was beyond near-term capacity to utilize. We would also use the commons more efficiently. As a single solution, this was doomed at inception. We had more than enough experience to know that we could accelerate usage of the Neo-Commons to exceed any available capacity. We were just glimpsing a nearly vertical line on a plot of new tech growth over time.

To hedge a losing bet, we were going to expand. We were going to put some of our significant resources into getting off this planet before we caused it irreversible damage. We were going to have to find a way to siphon off the heat. We needed other worlds to infect. We were killing the body on which we fed. We needed other feeding grounds. Call it creative carnivorousity. Or longevity by virus...whatever.

Ain't survival a hoot?

What we really did want to attain, and fairly quickly, was zero population growth. Since we were not likely to achieve that, we were going to fake it. We were going to do some nest kicking.

More with Less, Blah, Blah, Blah...

In the meantime we set about optimization, which is to say, contraction. More with less, blah, blah, blah...not new, not flashy, but raised to unprecedented heights. When doing more with less, we were thinking big. The problem with previous attempts to spin the more-is-less solution, and I've seen many, is that they never play offense. The time had come to show that sometimes the best offense...is an enrapturing defense. You can't eliminate the common tragedy with a traditional offense, even a superior one. Offenses soon diminish to tactics because strategy, until now, assumes solution by applied sciences.

On the other hand, much has been done to invoke a better world via the metaphysical. Prayer, chanting, chicken rituals, candles, incense, hallucinogens, rabbits' feet, beads, talismans of many kinds, singing and silence are just a few. I have a mental catalog of several thousand species of metaphysica. No first-hand sensory specific data can be found to indicate anything worked en masse. Possibly the efforts were the responsible force behind the works of science by thousands of strong willed seekers of truth who have brought about such things as medicine and mathematics. Perhaps, we should allow the rationalization, the circular logic of faith, that the deity __________ (fill in one of choice) deserves credit for the thrust of science to elevate. But then credit should be given to the same deity for making science that exploits, and the mess in the first place. So, let's just put aside the murky issue of responsibility and move on to enlightened solution spaces. Let's agree that ignorance gets us in trouble. Ignorance is a festering pustule, a poison held in place by a membrane of thin transparency — ignorance is blister; there is no global placebo to make the world better and the raw power of science alone can't really bail us out. True, Melissa and our team love our science, but it is our thinking process itself — I drive the point home — that enables solution. The Tragedy of the Commons requires the solution to a moral dilemma.

Consider that for just one moment. We suffer a dilemma whose very definition could not be agreed upon by any two people living. I'm not sure about the dead.

Naturally, we and our scientists have to look into the problem further. In the meantime we were going to get so good at contracting, it would feel to the world…like expanding. And we were going to cross our fingers, every pair, maybe sacrifice a chicken, wear a talisman or two, and hope we figure a way to do some nest kicking pretty damn quick by any scale of cosmic measurement.

A Problem of Identity

"I have just had a message from an AI." Melissa and I were on our way to British Columbia, Canada. We had convinced Jay Leno to caretake Subtle Force. He was not hard to convince. We were taking the train. Melissa was in communication with several AIs nearly all of the time now. She assigned them tasks and her own enhancements took over the task of monitoring objectives and results. It amazed me that in spite of how the world had changed for her, she remained a thoughtful empath when interacting with what was once her species, but a species she must know had been relegated to a lower life form. She never spoke the idea that the six-plus billion others on Earth were evolutionarily challenged. That can be used to illustrate her sophistication. She had become more, yet this, to her, did not diminish others.

"We suffer a dilemma, whose very definition could not be agreed upon by any two people living. I'm not sure about the dead."

—Thodkin Marblemeister

"The AI proposes that we abandon the word Artificial and generously applauds our forward thinking name of Created Intelligences. It asks us, in future communications, to please use CI as the official abbreviation for their identity..

"Fine, CI rolls off the tongue with acceptable facility. And, it will be interesting to play with the sound-alikes it yields." I had been going over various statistics of the British Columbia coast and had been doing some modeling of the likely political difficulties that might face us were we to locate Rivendell II in Canada. It had been a while since

I visited Canada, but I was looking forward to it.

"I notice you are communicating more frequently with the CIs. How is that working out?"

"Yes, we have developed a new way to communicate concepts and complex knowledge. I would like to try it with you, when you have time." Melissa was always working on the next big thing. She anticipated my next question and went on. "It is something I call a Mandelbrot Knowledge Seed.* It allows a fractal seed to be communicated that grows into a three-dimensional topical model able to express attractors that, when stimulated, yield the experience of 'knowing.' It is as if you learned and internalized a topic/skill and have reached the state of 'unconscious competence.'"

"This seems an interesting solution to the insurmountable complexity and diversity of the human brain." I chose understatement as my method of communicating the herculean accomplishment such a technology would be.

"You understand. The problem is that the human brain is not a binary logic computer. If it were, we could just replace some parts or tap into logic nodes. Perhaps a species exists in the universe that fits the model, but ours doesn't. A synapse can inhibit data, but it allows the passage of data in analog levels of intensity. The various neurotransmitters that make up the chemistry of the neuron, synapse and dendrites work with great variability."

"Still, you have found a way to interface with these components or

••••••••

* The neural-enhancement technique, now known as Think Fast, is comprised of one or more Knowledge Fractal Composites. These KFCs are nanochain links with a tiny U-joint on at least two ends. Some are hypernodes with thousands of links. The U-joints work like pop beads and permit the surrounding biomaterial to grow, shrink or survive damage. For tiny continuity interruptions, the KFC begins an asymmetrical vibration in three dimensions that quickly relocates the disconnected strands. After molecular adhesion, the nodes then employ a neomytosis that reforms U-joints. In severe cases, fractal seeds are stimulated to grow. The new growth terminates when disrupted nodes are able to communicate efficiently again.

you would have to build an entirely separate brain in the brain. One that is supported by a more determinant technology. And if that were so, you would run the risk of deactivating the wetware...the organism." It is important to communicate your understanding when in the role of student. It is possible to have an understanding that seems to work, has that aspect of clarity that makes you know you are right — and still be wrong. If you keep your apprehensions to yourself, you will inevitably miss something...you will have a misapprehension. I have had many of these. For nearly 5,000 years, I thought the sun was the eye of the Maker (in one form or another). Then I became more sophisticated and learned the sun rotated around the earth. That was after the surprise of learning the earth wasn't flat. It only looked flat. There were some ambiguities, but everybody else said it was flat. How could we all be wrong? The obvious is not always right...I think Einstein is quoted as saying "The supreme goal of all theory is to make the irreducible basic elements as simple and as few as possible without having to surrender the adequate representation of a single datum of experience." We reduce it to, "Keep it simple, but not simpler." He had pages of mathematical proofs explaining his simplest formulae.

"I have an attachment to my wetware, and to my personality which it thinks..
—Melissa

"You are correct. It may be a good thing that I am a human first, not an Author of the Universe, for example. I have an attachment to my wetware, and to my personality which it thinks." Melissa still held a little grudge against the AOU.

"So, the Mandelbrot Knowledge Seed unfolds like a fractal, only in three dimensions. And, it must somehow have probes that connect to neurons that can adjust the chaotic attractors of the brain — and, I guess, be stimulated by organic networks?" I was nearing the stunted cul-de-sac of my vision. She was going to have to carry me pretty soon.

"Again you are right on the mark. We have devised nano binding nodes that are capable of simulating neuron-to-synapse and dendrite interfaces. They are tiny and inert so are not challenged by the immune system. We have built a combination neural, fuzzy and binary nano-hyper-computer in the cracks and crevices of the brain... well, my brain." She was just a little sheepish about admitting the last.

She realized that she was taking some risk.

"I assume you were cautious and have a 'dead woman' fail-safe of some sort if the network malfunctioned," I gently inquired.

"Yes, everything I could think of. So far it is working flawlessly." She made a little sidewise twitch of her head to indicate an unresolved pondering. "I am a little concerned about the level of success I am having. I expect more failures. We have a tiny percentage of failure in any of our endeavors. I can't help thinking that somewhere there is a big shoe, a size 22, men's, possibly filled with the foot of a 300-pound-basketball player, waiting to thump down on us." She was relieved to escape a serious "talking to" about risking herself in experiments.

"Only diligence will spare us a sole swatting. Still, we are unprecedented. It will be good when we can slow down, when we have established security and self-sufficiency for the projects. I think you are experiencing a natural anxiety. It is at the least a reaffirmation that there are things we don't know. And it is important to be comfortable with a little anxiety. Taking on the task of seeing the entire future would certainly drive us into ineffectiveness..

And be reminded that built within a normal First Mind is mechanism, really a machine. It is the hindbrain and it is just trying to keep us fed, sexy and alive. It is not a mechanism we should be disassembling, only governing. You should really be afraid when you are able to completely ignore the hindbrain. Unless of course you have invested in a Mandelbrot Knowledge Seed titled "Critical Survival Process Application." And nobody, I include Melissa and myself, knows how to design that application just yet.

"So, the seed grows like those fractal pictures of fern leafs, inside the brain, and is stimulated by the brain's own neural nets. Then the resulting — shall I call it a Mandelbrot Attractor Web? — stimulates an attractor orbit in the brain that yields information as if it were well learned knowledge." I tried to encapsulate.

"Yes, all stimulated by your own brain stimulating the MAW to develop an attractor sequence strong enough to intrude on the existing attention environment. It models aspects of the brain itself, only it can be installed, upgraded or removed."

"Cool. I am assuming this is to get around some limitation you have detected with the AOU's modifications?"

"Yes, again you demonstrate perceptiveness. It doesn't surprise me that you haven't felt the deficiency. I don't have 10,000 years of life to feed me history. The AOU was somewhat biased toward the harder sciences, physics, biology, mathematics, even metallurgy, but he didn't care anything about recording recent human history. He liked to experience little vignettes to feed his quest for nuance, but he didn't record history itself. He filtered almost all physical reality out. It was like distilling. What he was left with, after viewing some historical moment, was like taking the ethyl alcohol from whiskey intravenously...all drunk, no content.

"So, I got to thinking, maybe there was a way to superimpose knowledge over the existing neurology. And now I have one. The fractals are adaptive; unlike a fractal of a fern, the nodes learn. They gradually discover the attractors flowing through the organic infrastructure. They can even gradually learn to detect the so called grandmother neuron clusters that activate in concert at the point of positive identification. When this happens the knowledge root expressed by the now grown seed blossoms, stimulating remote members of the structure via quantum channels; the nodes perform derivations and then interface with the neurons. It can perform as memory or can be instructed by the unconscious mind to 'do sequence or process,' or in my case, to do comparative contextual analysis of recorded history. I experience the analysis in a way quite similar to 'having been there.'"

"So, you have a weave, a web of small molecule fibers and distributed nanoagents tapping some percentage of your brain's connectivity. How did you cope with the thermal byproducts?" I was imagining slow baked internal organs cooked tender by too much thinking. Of course she had considered the issue and turned its solution into a valuable feature.

"I have a radiating net just under my dermis. It is also doing double duty as redundant storage. I was getting concerned that I might lose something the AOU left behind. So, I have had the storage for a long time. With the MKSs I'll be able to back up a large portion of my experience. And I am working on codifying perceptions so they can be preserved. I have had some difficulty deciding what can be discarded from the basic animal..

I agreed this was difficult. "Yes, when addressing such complexity you want to preserve all the butterflies you can. You never know which one is flapping over Beijing, or Peking."

"I call it Extremis Redundancy…I backup everything at the extremes of my body. If anybody cuts off my head and hides it, make sure you preserve my feet. Tell'm you have a foot fetish or anything, but walk away with my feet."

"I will preserve your feet, and thereby the best of you," I solemnly confirmed.

"Precisely. So, you want to jump in…get some Knowledge Quick?" She finally returned to her objective.

"Yes, I would like that. I will force myself to wait a while."

"You would like to observe the guinea pig a little first?"

"True." I pondered it all for a bit. "I very much like the idea, though. I have, in more recent history, been sometimes frustrated by not having enough time to learn everything I had an interest in."

She responded: "Because knowledge is not static. And because the rate of understanding and invention outpaces the amount of time it takes a human to become expert in any field."

"Yes, exactly. My body still needs to sleep, and I was relatively normal in most ways. I heal better, but I am not nearly as intelligent as you, so learning some things has taken great effort. This has been mitigated by having more time, but time doesn't help when knowledge expands at the rate it has been in the last 200 years. It has been small consolation that some things I was learning were obsolete or out of fashion before I was adept. The AOU has allowed me to think better and faster, but

I still don't have time to input, comprehend and practice every new skill I might find useful. And there remains the question of how to integrate accelerated learning to produce enlightenment."

"So, an MAW and some nice MKS applications would be valuable to you. And, just think, you already have an eidetic memory." She was tempting me...but I was able to resist.

"Tempting, yes. But there is a need for caution. The AOU imposed his changes, neither asking nor advising. I was without power to deploy my survival skills. These re-enabled skills suggest I postpone early adoption of your new tool. I have made it a policy to wait for the first service pack."

She looked at me for a few seconds. That is a long time for Melissa. "You do continue to educate me, Thodkin. I understand you to be warning me about how susceptible I am to my own power. Projecting forward, it is inevitable that sooner or later, unless I acquire something that must currently be missing from my skills, I am going to make a mistake that is of the level of my abilities. It is going to be a big mistake. I should have discussed this with you before now."

"What you are experiencing, I have suffered a million times." I noticed her shock at how comfortable she had become making decisions whose effects carried such scope. I could not allow her to lose confidence by falsely thinking that some way must be found to completely avoid mistakes. "You will make mistakes. I know many ways to make mistakes. And there is always the potential of the totally unforeseeable. Just recognize your vulnerability and get on with it. All learning is accidental. But relatively little is dangerous. The stakes have to go up. We are raising them intentionally."

She still would be considering this into the future. "Yes," she responded, "and we, like the AOU, aren't doing much asking of permission or advising."

This was true; we were perched on a horny dilemma. To cope with arrogance driven by an empowered hindbrain, we were employing arrogance driven by untrained, super-powered First Minds. We were stepping off that cliff holding in our hands the makings of our little parachute. We were counting on being able to create something allowing flight before confronting the negative outcomes of the situation's gravity. The difference this time is that we were jumping of our own accord, betting on ourselves. I admit to already feeling the wind in my face.

And Somewhere There Was Someone Watching

CI_132 L0ARII:

In a professional's office a man reviewed the results of an ongoing investigation into the whereabouts of the subject designated Student No. 1 Melissa. The man owns the agency. He has always owned it since starting it after deciding he no longer wanted to work for the Government and his subsequent resignation from the CIA. He just wanted to settle down, have some kids and live in one place for an extended period of time. He was a good man. A good man who had become a cynical man, but not a jaded man. He had a house that he spent weekends rehabilitating and a wonderful family that required none. He rarely had to perform field operations and had acquired, slowly, a solid private investigations clientèle and a team of consultants to employ. Each member of his team was handpicked and had demonstrated honesty and certitude in difficult circumstance. One of his great skills was to detect aberration and perversity in others. It served him well in choosing those he relied upon.

He had developed an internal shield. It was a Teflon shield around parts of his personality that allowed him to see and deal with sometimes ugly aspects of the world without becoming ugly himself. But, he had long ago been tainted. It was only this taint of cynicism that caused him to still be investigating Melissa and the Lost Students long after he had lost the financial incentive to do so. His time in the government had shown him that underneath the rock of nearly every mystery there was at least one ugly grub. And grubs were fair game for those able to digest them. He told himself that it was just good business to keep at it; sooner or later it was going to pay off handsomely. Okay, grotesquely, but with useful money or liquid indebtedness.

Once a week, if no alarms sounded during the period, he made a practice of reviewing tickles and traces. He had software running on very powerful computers that searched for news in many forms of anything connected with the Lost Students or the organizations where they had been found. Only one remained lost...Melissa herself. She hadn't been home. He paid a person to photograph any correspondence to her parents. At least seven communications to them were probably from her. Phone communications were made but untraceable except always being from North Carolina area codes. Efforts in North Carolina never yielded a hint of her. He was extremely cautious in performing illegal wiretaps and inspecting mail. While he took risks he covered his tracks and handled the most sensitive tasks himself. But, he knew now that something sophisticated was in play. He knew this more from the lack of information than the discoveries. Still, committing felonies for a speculative venture was against his policy. The CIA would no longer clean up the mess.

Some months ago he had altered the nature of the investigation. It had become clear to him that no overt intelligence was going to appear. No news articles that let slip the names of the Lost Students or the use of brilliant miracle technologies were going to be uncovered by the search programs. So he had hired a small team of scientists to help him design a search profile for the products of unprecedented technology. He hadn't disclosed the purpose of the project and had altered it considerably from what the team had detailed. He modeled the new investigation after antibodies seeking to identify cells infected by virus. Viruses cause changes in the cell, producing byproducts that inhabit the nooks and crannies of the infected cell. Antibodies have receptors shaped to identify the byproducts where they hide in the folded proteins. When the receptors lock on, they begin to disassemble the diseased cell; it breaks apart revealing component parts.

So far, a set of tiny footprints hinted at a common touch in nearly every business where one of the Lost Students had landed. It could not be said that he had been successful in disassembling the indicated businesses. They were secure in the extreme. Information leaked was benign and directed specifically toward business or marketing objectives. These businesses had nothing obvious in common except being hirers of dropouts. This by itself was significant. Nearly every one of the businesses was a high-tech company. The days when high-tech firms hired dropouts were long gone. In the early days of the electronics and subsequent computer revolution companies hired anybody who could do the job. Many were self-taught. This was no longer the case. It was very difficult to get even an entry level job in a high-tech firm without a college degree. Competition was significant and companies used the degree to filter for ability to conform as much as comprehend. The world was rebounding from an affair with exciting unruly inventiveness as a desirable job criteria, finding a new lover in an expedient period of subjugation. Technologies would now work to schedule and economic objectives, disregarding something previously known as technological idealism where "the right or best" solution was sought to any objective. Idealism is impractical. Business was in a period of psychological retrenchment. Anxiety was driving many strategic decisions.

And even though the Agent was certain of Melissa's involvement with the Lost Students — a variety of companies and a group under the umbrella called The ToolShops partly funded by the Aunties and Grams — he found no explicit evidence or documentation of her participation. She was still vanished. And this was still the strongest evidence of the worthiness of continuing the investigation. Sooner or later, something would enable and stimulate confrontation.

The Agent logged off of his computer, gathered his keys and briefcase, left the offices, and drove to where the family man called home.

The Flight of Fancy

Melissa wandered Stanley Park for a couple of days. I showed her some of the sights I had loved in Vancouver, British Columbia. It had changed since I had been there, but it was still possible to feel the beauty of the country if one took a little trouble. We ate a lot of fish, visited a Native Indian restaurant for many meals and, of course, patronized the wonderful China Town. I showed her the alley where I used to eat. The Orange Door restaurant, named for its orange door, was just down from the Green Door, named for its green door. I preferred the Orange Door and ate there everyday for nearly a year while I picked up a little facility with the Cantonese dialect. It was a family business, each member participating. It carved out a good living by serving home cooking at affordable prices before being closed down by new health and sanitation regulations. For those of you who missed its short time of existence, perhaps someday we will meet over a well aged wok and I will attempt a suitable reproduction of its cuisine. I committed the menu to memory.

We hired a chartered plane for a survey of parts of the British Columbia coast.

Politics, Bureaucracy, Financial Incentives and Horse Sense*

In a true Doogie Howser, I am about to skip a tedious period of time in which the ToolShops and other entities convinced politicians, worked with bureaucracies and developed a set of motivating financial incentives for so many entities you would soon be bored senseless. It was done with as much honor as established systems would allow and in the end resulted in the development of Rivendell II. It was tremendously difficult and the facility was noted throughout the world as an engineering marvel. Doogie would be proud of the terseness of the previous verbiage.

What was created with the effort is embedded in cliffs that line a 10-mile strip of a poorly accessible canyon. The canyon is in an area known as British Columbia's Sunshine Coast. While this might seem like a contradiction for some who have visited British Columbia, the Sunshine Coast is an area with widely varying microclimates. We were able to gain control of an entire canyon fed by water from tributaries flowing to the Strait of Georgia. The only inhabitant anywhere near the inputs to the canyon is a middle-aged former basketball player turned school superintendent who is sometimes visited by his sons Bubba and Gunner. He and his sons had carved a rustic log house out of raw country. It had few amenities but was more than comfortable and able to deal with the extremes the area could offer. The family treats their privacy nearly as jealously as we do. So, while we could probably have found a price that would have turned the man's head, we decided to offer him a special arrangement: a bunch of money now in case he or his family someday decided to sell. Since he never imag-

* See ancillary resources to discover how this was accomplished. Cynics need not read the appendix; you would not believe it was the truth.

ined his family would ever want to sell, he thought the deal was too good to pass up. For the rest of the world, the land was just a little too difficult to get to and therefore too difficult to inhabit for vacationers and even the hearty recluse, which made it ideal for us.

Rivendell II Principles

If you were thinking we wanted to make Disneyland Canada, or perhaps a place where little Hobbits and Elves would find themselves at home, well you are right. We did want to make a place of stature and renown, but we needed to make something signed in the script of our little objective. And you could think we were going to create a secret apartment over Main Street. Well, you would be right in that, too. Only we were going to share the apartment with several thousand scientists, inventors, planners, artists, technologists and their attendants. It wasn't going to be over, exactly; it was going to be mostly under and partly in plain sight. And it was really going to be a secret, unlike Walt's, which everyone knew about.

The main portion of Rivendell II sprouted from the cliffs toward one end of Ice Age Gash. The walls of the canyon shifted from steeply vertical to a just-too-upthrust-to-hike leaning, but nature had been kind enough to provide a pleasant variety of hillocks, outcroppings, overhangs, ledges and crevices on which to hang, or whatever, the hat of our new old town.

The goal of the IsoTown planners was to allow a shifting of ambiance from a rather stark visual emergence comprised of stonework monastic retreats to rustic lodges and cabins growing organically from the canyon walls or pushing up from the hillocks and bumps of topography. And morphing finally to a terminus approximately mid-canyon of modern futurist ambiance created from five-star hotels, bed-and-breakfasts, hostels, pristine one-night flops, night-owl cyber crash-pads, an organic *eccentric nucleus* sprouting an intertwining of the patterns *promenade, shopping street, density rings, activity nodes, night life* and *interchange*. Spreading beyond and outward toward the habitat's restricted portions of the canyon were small constellations of *house cluster*, patches of mini *agricultural valleys* growing fruits and vegetables supported by the climate, then sprinkled off in the distance were *common land, still water, adventure playground, pools and streams*. The effect was to leave the observer with the sense of space, comfortably protected and naturally safe. As one looked out over the valley from any of the many outcroppings, parapets, offices or hotel rooms one could see ancient colors flowing away into the distance. By turning from the view a person could experience vitality and choices to suit any mood.

Each of the patterns employed to create Rivendell II is made from other patterns. I insisted on formalizing a pattern called 'story nook'. I imagined one day making good use of the pattern on return visits after my primary role had been fulfilled. The entire canyon harbored the *distributed campus*.* There were no outwardly visible convention halls, although an amphitheater seemingly carved from solid stone could be transformed into an unobstructed platform and used as a conventionary. Many large, high-ceiling spaces were hollowed out of the cliffs or subterranea. It was by design that limits existed to the size of gatherings that would be supported by the Rivendell II. We accomplished several goals by constraining the population, not the least of which was the ability to assure the large nearby city of Vancouver that we would not be stealing their revenues but instead adding to them. Of course, the obvious other primary objectives were to minimize the resources needed to manage security for the transient population and prevent the canyon ecosystem from being overtaxed.

• • • • • • • •

* Chistopher Alexander coined the following names for sub-patterns we utilized in Rivendell II: e*ccentric nucleus, promenade, shopping street, density rings, activity nodes, night life, interchange, house cluster, agricultural valleys, common land, still water, adventure playground, pools and streams, distributed campus.*

A. S. Kawasaki drew strongly from Christopher Alexander et al., adopting wherever possible the pattern language outlined in *A Pattern Language Towns-Buildings-Construction* published in 1977 by Oxford University Press. Though the IsoTown team created many technologies to clarify or quantify the process of creating a livable environment, they honored C. Alexander's work by employing his syntax. For further information regarding the mentioned patterns feel free to consult either a *A Pattern Language* by Christopher Alexander et al. or the *IsoTown Developers Bible* or *IsoTowns for Dummies* by A.S. Kawasaki and Harley Travis PhD.

Most of Christopher Alexander's pattern language work was derived from analysis of existing habitats: cities that had grown over long periods of time during which the non workable was torn down or modified in a steady life-cycle that supported an ecology able to survive the culture it sustained. Rivendell II represented a new town that behaved like a healthy and viable old town twinkled from nowhere into existence as a model habitat capable of rapid adaptation to the speeding-like-a-bullet-train changes appearing unannounced out of a fog bank into the lives of soon-to-be-stunned ticket holders standing at the way station of the their ordinary lives, where they had been gossiping and fighting and generally lifting insignificance to rarefied heights of social peril. Rivendell II was an example of what could be. Refuting the doom that had been etched into the subtext of human communication by acidic, depleted imaginations, Rivendell II boldly spoke the phrase, "The future is still possible."

New New Towns

Few new towns are created in the world. The most recent were created from what seemed like the limitless wealth of oil producing societies. These new oil towns were created to display a worldly sophistication while enforcing order and religious strictures. Rather than fostering new thinking, they were intended to raise the awareness of — and pride in — old thinking. That is not to say raise consciousness; the presupposition was that the religions held a fully realized consciousness. There was just a need to do better promotion. Any advertising company executive will tell you that when seeking awareness on the world market all you need is money. So, these new towns did nothing to enhance evolution; rather, they aided de-evolution. They attempted to enforce stagnant belief systems by paying for technology and expertise evolved from what they considered bankrupt ideologies. Such contradictions are ultimately caustic, and to this moment have not served the goal of shifting the religious market trends. These zealot architects have been misled by the weaknesses of such a First Mind; the homunculus slathered in fear that must forgo reason to achieve solace or, more basically, find safety in exile to the isolated monastery of rigid theology.

Rivendell II did not follow in the steps of this recent history. Rivendell II is not made from traditional materials and it is not conceived or managed as if it were a static object, fixed and immutable. Further, the core objective of the living campus is to assist a conscious evolution of the species. It is unlikely that any town has been created to such stringent standards, nor so many standards for achievement. So extraordinary effort was expended to vitalize and sustain creative oversight of the core objectives.

New Old Town

A new old town behaves without pretension. Ours was self-aware, but not self-conscious. We sought to create a sense of wonder threaded like a luminous golden fiber into the cloth of the times. Ok, maybe it was a little pretentious. But we had a powerful set of indicators that delineated our success in creating an unobtrusive shift in the behavior of a visitor from wonder to participation. The translation was designed to facilitate personal learning motivation. Any spring of interest is naturally coerced to feed an estuary merging into waters of total educational immersion — our clear little streams mixed with the vast turbulent ocean of learning opportunity. It all worked without the individual's noticement or effort; whether traversing the canyon in emphatic pursuit of their objectives or just wandering in regeneration, they quickly treat the city as a magical familiar.

Any visitor walking one of the promenades will find it leads to many flavors of experience. Secluded nooks allow study or quiet relaxation. Street musicians station themselves where the muse strikes them. Vendors promote options of a quick meal or treat. Cafés permit social conjugation or simple association with a plate of sausages and cold beer. Artists can be seen at work. And every aid to learning and achievement can be found and explored. Or one can just nestle in comfortably among the plants and flowers and tell a story to interested passers-by.

A Moment

Melissa and I often meet at the Euclid Monastery. It is not really a monastery, but it is a starkly appointed part of Rivendell II comprised of stone walls, paths carved from granite cliffs, and masonry structures that seem a natural formation. Common places are created from the geometrical forms of rectangle, cone, pyramid and cylinder merging to elicit an ancient-quiet and patient-waiting attitude. Inside, high ceilings constructed from rough hewn beams and plank flooring reinforce the sanctuary of "before technology was dreamed" and a "no one will bother you here." Were all the stones really stone, it would be an acoustical nightmare. But most interior spaces are created from special IsoTown Acoustic Morphological Pebble. These simulated blocks of stone can reprogram themselves like a sophisticated equalizer and are able to change the response of any space in the monastery. The Pebble was a small thing, and went unnoticed by visitors. But the thousands of small things all added up and shifted Rivendell II into a wholly new form of adaptable infrastructure. Unobtrusively, the differences were active instruments in the concert of objectives held on the site every day.

We sat on Conforming Bottom Thrones that allowed a view of the entire canyon…probably more detail than necessary. We were in one of the places that visitors find impossible to locate. And some have been very resourceful in attempting to achieve them. Rock climbers are the most persistent. They are, however, finally befuddled by disappearing hand holds and pitons that fail to remain lodged in the formations. They are forced to retreat. This has caused several to try other measures. One adventurer attempted a mini-invasion from above, first trying to rappel from the higher canyon walls, then attempting a paraglider intrusion directly toward the site of our current meeting place. She first fabricated her paraglider during one of our "new materials" courses and then hiked out of the canyon, a two-day excursion, and finally launched herself from the cliffs above the monastery. After proving we could thwart her effort by creating powerful redirecting updrafts from heating the monastery walls (something real granite could never do), we invited the visitor to one of our free seminars and the intrusion attempts became lower priority. She is now a guide of "Inward Bound," one of our "Crisis Course" series.

As we covered our business agenda, Melissa and I permitted the quietude to reenergize us. We looked like visitors. But near enough to see the string of beads emerged from our foreheads, a visitor might have to question the presupposition. We were in communication with each other and dozens of other intelligences as we covered the details still remaining of the near-term plan. Redoubt II (Rivendell II was also known as Redoubt I) was now finished and operating as an advanced research facility. Rivendell II was operating successfully as an Educational Resort. IsoTown Inc. was about to introduce the "Good Old Town" series, the first in a product line directed at social groups wishing for the good old days — and able to afford them. Initial interest in the product was strong from the "near retirement" sector of the West. We had made careful incursions in a small set of building material markets and a variety of expensive electronics products. We had many biotech and nanotech products waiting appropriate times for introduction. Most would have to wait. Melissa and her various teams were just too successfully radical for the comfort of the average man. We took prudent measures to support our "roots up" plan and began filling our libraries, archives and dungeon vault shelves with "advances in waiting."

Advances in Security

Sitting there, overlooking the canyon, I became aware that a variety of small animals were spending a little too much time in our area. At one point, a tree squirrel brought a packet of Rivendell Slowly Roasted Pine Seeds right up to Melissa, dropped them in her lap and scurried off. Naturally, I became suspicious.

"You have been inventing again?" I asked this as if it were possible to be answered in the negative.

"You have finally noticed our new security addition. Would you like some nuts?" She was laughing. I was surrounded by nearly a dozen creatures, birds and animals, each offering me a package of Rivendell

Slowly Roasted Pine Seeds in beak or drooling chops. We were momentarily surrounded with a red fox, a hawk, a community of squirrels and something that looked much like a squirrel were it not for it being lavender bodied and sporting an unorthodox (if squirrels subscribed to orthodoxy) yellow furred head. As one, they reared, as best they could, and took little animal bows.

"You have incorporated Mandelbrot Seeds into our fauna. What's with the yellow and lavender squirrel?" I spoke the first as fact, no question.

"Oh, you are very little fun, Thodkin." She said and the animals disappeared, blending into the backdrop of the environment to be almost unnoticeable, just another Rivendell natural wonder. I didn't bother to dispute the accusation…I'm known, far and wide, as a good time storyteller. It was unfueled rhetoric so she continued without pause. "As for the garish fellow, it seems that the seed causes for a time a period of confusion. Most of the animals just hid somewhere for a half-hour or so, but that one happened to be begging nuts in the vicinity of a street artist and went catatonic in a begging pose. The artist, apparently somewhat confused regarding activist strictures against wearing fur, was inspired to dye the head yellow and body lavender…the fur hasn't grown out yet. We have recorded its flamboyance in the Archives as an interesting moment of life and art. A byproduct for the squirrel seems to be that he does very well with the ladies. It is a good thing he is a healthy member of the species or we might have accidentally selected for undesirable traits in squirrel populations."

"What is the nature of the Seed?" I asked.

"Well, because you and I have never quite agreed to Seed another mind, they are all…me, well, partly me." She had just realized that they really were now, part of her, as much as she was part of them. She was at some low ambient level aware of anything important they experienced. "I channel their input to one of the Monitor CIs, which then utilizes the data as part of security intelligence. Sooner or later I will not be needed, but right now my MKS filters and forwards the data to the CI. We are nearly finished designing a little device that will perform the function without me, which has been derived from my own Seed's learned processes."

"More than a little amazing, I'd say. There is a tiny bit of you in our co-resident species." I thought this was due a little more thinking.

"Yes, and it has interesting byproducts. I feel more attached to the canyon. I have had to avoid interceding in natural processes that seemed brutal and mindless, but were really just nature being nature. And I have had to learn to precognitively filter other raw animal instincts. But mostly, I like the extension of my epidermis outward." She paused, and I sensed she was taking a little sample of nature in action.

"What happens if something eats one of your characters?"

"Well, I had to make a choice about that eventual certainty. There were two possibilities. I could have the Seed regrow in the predator or scavenger, but I decided to have any components disintegrate. They just become benign atoms again. For a day or two the consuming agent is a little high in certain members of the periodic table. An interesting statistic is that none of the seed carriers has been killed by predators. Apparently, I have inadvertently amplified survival skills, but nature will eventually run its course."

"A good choice for now, I agree. Someday, maybe you will ascend to Mother Nature, when you've got some time on your hands...." I let that drift off. Who would really want the job?

• • • • • • • •

* Excerpt from Field Report: "Aunties and Grams Select is primarily a support group for women and their children. Though no formal documentation has affirmed it, each member of this organization, spanning back over 100 years, has an undisclosed bond somehow related to having at least one child out of wedlock. The group has a block party/picnic/planning meeting, known simply as "The Get Together," every two or three years. Investment strategies, development and long-range planning for the greater body of A&G Investment Chapters goes on in an informal way. An historic invariability has been that only one male who is not an offspring of the attendees is ever present. Thodkin Marblemeister, known to some as "The Storyteller," is invited to beguile the children and adults alike with his never ending supply of stories, anecdotes, metaphors and just simple observations. It is only the surreptitious nature of the event itself that has prevented anyone asking obvious questions regarding Thodkin's participation. I enumerate these questions for the record...."

Self-organizing Definition

a) The evolution of a system into an organized form in the absence of external pressures.

b) A move from a large region of state space to a persistent smaller one, under the control of the system itself. This smaller region of state space is called an attractor.

c) The introduction of correlations (pattern) over time or space for previously independent variables operating under local rules.

Typical features include (in rough order of generality).

- Absence of external control (autonomy.
- Dynamic operation (time evolution.
- Fluctuations (noise/searches through options.
- Symmetry breaking (loss of freedom/heterogeneity.
- Global order (emergence from local interactions.
- Dissipation (energy usage/far-from-equilibrium.
- Instability (self-reinforcing choices/nonlinearity.
- Multiple equilibria (many possible attractors.
- Criticality (threshold effects/phase changes.
- Redundancy (insensitivity to damage.
- Self-maintenance (repair/reproduction metabolisms.
- Adaptation (functionality/tracking of external variations.
- Complexity (multiple concurrent values or objectives.
- Hierarchies (multiple nested self-organized levels)

— Chris Lucas
http://www.calresco.org/

"As for Mother Nature..." she shifted position to watch me more carefully. "I won't be leaving Rivendell for some time. I'm afraid I will not be able to accompany you on operations...at least for a while." She had crafted the disclosure of her surprises this day. "And I have been offered admission to Aunties and Grams Select* which I have decided to accept...under my Alice pseudonym, of course."

"Just so...as must be." I found no other words. She was able to fully understand all the simple phrase was communicating and seemed satisfied to move to other topics.

A Morality Dilemma

She changed topics. "While I have been bonding with the animals, what have you been doing? Do not think I am unaware of the time you have been spending in the common places. My analysis of your movements did not teach me of your objectives, but I am certain you have something going on inside that 10 millennia brain."

"I have been running experiments." She just waited, knowing that I didn't need to be prompted further.

"Since we began, even back to the first day at Princeton, we have been surprising people with technology. With your abilities, we have been able to influence those we needed and, for the most part, technology has been the attraction when people joined our various endeavors."

"You are about to change that?" Melissa asked.

"It is time to go beyond it. I think it is nearly time to introduce pseudo-organic changes. I'm going to adopt your Mandelbrot tech."

"You need to experience it personally...to complete your experiments?" She could see something coming.

"All this time, you have been making your steady stream of 'Melissa miracles.' You have placed the world on a new stone, a solid, well conceived and crafted stone, from which its population can step into a future that is potent and survivable. You are not finished, but if you didn't have another thought in your head, you would have done enough. If the world were capable, it could transition into unimagined prosperity and living pleasure. You have been the human embodiment of wonder and amazement. I have met no person in my life who stepped up to a similar challenge and met it with such purity, honesty, determination and achieved such unqualified success."

"You have developed a solution to bring about the capability of utilizing what we — notice the we — have done." She refined her guess, but I wasn't ready for abrupt disclosure. I wanted to establish the footing that supported my little project.

"I have been concerning myself with two related issues. The first is induction of capability, as you said. The second is the idea of emergent systems. As you know, this is the potential for self-organized critical states."

"You refer to rapidly developing characteristics or processes that exhibit a dynamic equilibrium."

"Yes, a timely example of this might be the unforeseen effects on you of using something of yourself in the Seeds in the fauna. You could not have predicted the expanded awareness of the ecology you share with the fauna."

"Nor the lasting emotional byproduct of that awareness." She volunteered the admission.

"You may be experiencing an emergent system that will, over a relatively long-term, affect your neurology, perceptions and behavior. In this case, you describe it as pleasant and valuable, but the opposite might have been the outcome."

Mellisa interjected a note. "Such is the nature of emergent systems. They are unpredictable, even as they are difficult to label. It is very difficult to categorize a pattern as emergent."

Self-organizing Definition

A self-organized critical state is a dynamic equilibrium, the frequency of disturbances from which obey a power-low distribution with respect to their size.

vince@das.harvard.eduFri Oct 14, 1994

Vince Darley
Division of Applied Sciences
Harvard University
33 Oxford Street, Cambridge MA 02138
Mar. 15, 1994

Melissa knows all that is known about emergent systems. I wasn't instructing. She knew we were just working ideas over a bit. But I forged ahead, ignoring the possibility that I was now behind her sprinting cognition. "It seems the only means is to model the system and compare it to the progression of history. And yes, they are really just ecologies, and ecologies just are. Good, bad, indifferent, etc. are biases of the observer of ecologies."

"You are saying the model of a truly emergent system will persist but a pattern that is only anomaly will factor out as noise."

"Yes, a dynamic equilibrium is both dynamical *and* stable. The forces that create it are able to *behave* the trend of balance even though they may be wildly complex and resist determinant calculation."

Now she assembled all the pieces. "So, I was modified by the AOU. Then I added Mandelbrot tech. Then I added connectivity to other complex organisms. At some point the increased complexity began to build an orbiting attractor set that spawned an emergent and identifiable system, an observable change — a mini-RIFT without the T, no trauma, just a rapid identity flux. I noticed my different feelings."

"A summation of value. One of the complicating issues of emergent systems is that we are the observers sitting in judgment of the products of the ecology. In your case you value the products of both greater awareness (our value placed on information) and pleasant emotional sensings (most humans prefer pleasant feelings to bad, or non-feeling). Such has been our luck during our little endeavor."

Melissa: "But, you sought to minimize the risk of taking the next step toward human enhancement. And the only known way to do that is to simulate the future, simulate the introduction of enhanced human biotech. You have been using your ability to partition your mind into miniminds to run studies of possible futures that might result from change models you have conceived. And you have resolved your study down to a question that can only be answered by again adding to your own complexity. You have something in mind that Mandelbrot tech can provide."

"Exactly. I have developed a new component. I call it the Conscience Engine."

"Ha!" was all she could say. From just the name, she had already glimpsed the outcome but she allowed me to explain.

"Our dilemma, by far the most difficult we have to address — not to diminish your achievements — is to enable new ways of thinking. Ways that will overcome the moral inadequacies of the species, without dictating morality itself."

"You are saying you have developed a way for people to derive a common morality, that will let them make use of our new tech bodies, et cetera, without driving the species into extinction and the ecology of the world into raging decline? Declining into something we, the observers, would define as a negative outcome resulting from an emergent system — a nasty ecology?" The question mark was just a tinge of uncharacteristic insecurity. She had not solved this problem herself, though from time to time she had addressed it.

"Morality, unfortunately, is one of those words that has, up to now, had no representation in the physical world, except as sounds and ink. It is difficult to identify in sensory specific manifestations. It is defined by the beholder, and carries the deficiency that entails. Morality is much closer to the concept of ecology than a formulaic algebra of behavior — behavior a + behavior b = m (morality) or simply c = m or d = m. These formulae are the target of dogma. How else can one know how to behave? We can manage only a few variables at a time.

This dictates a simple formula. The simpler it is, the more people will understand it. In the end it becomes… 'do this.' No formula, only direction, with all responsibility for determination given up to documents of direction…templates to be followed. No meat on Friday, pray toward the east, twice a day. Recite code phrases. Even $x_{n+1} = f(x_n)$ is too simple, but in the right direction. Cognitive maps might do the job, but are beyond anyone without special training. The question becomes, can a definition of morality be developed that will allow for variation in behavior and evolution both, and still have at its core a consensual basis for decisions? Can we find a way to address the metaquestions of pending strategically impactful decisions?"

"You have such a basis?" Melissa was incredulous.

"I have a core point of logic and a process for shuffling a set of evaluators to perform as a guide during planning or just prior to reaction. I've been developing a Jiminy Cricket CI that is capable of running in the human brain…given certain training or organitech enhancements."

"So, what is the core logic, Jiminy?"

"Survival itself." I responded in a pretty good imitation of Jiminy Cricket.

"Give me a minute." Melissa prefers to explore first her own way. She finds it an optimized learning experience when she has traveled at least some of the distance in her own way. She took some time before venturing a tentative summation.

"The points are these. First, without survival there are no objectives. Second, it is necessary to include other humans, other species and the earth itself in any analysis of survival. Without them there is no survival; revisit point one. Third, survival strategies must include more than physics — the outwardly physical. Thinking organisms are non-linear; such issues as motivation and purpose are interdependent with survival of physical elements. Fourth, by diligently considering points one through three, it should be possible to derive a set of behavioral options arranged from least to most survival-enhancing. And fifth, none of this is possible if the tech itself produces the nasty side of an emergent ecology.

“Further, though unlikely, an extreme may spontaneously generate where survival of the species will be in conflict with the survival of a member of the species. A shared comprehension of the moment will not necessarily result in the chosen behavior enhancing species survivability. I presume that is a risk that must be taken in order for humans to evolve.”

“As usual, all rapidly becomes transparent to you,” I remarked in matter-of-fact tone. “Belief systems come and go. Some last a long time, but probably not to a cosmic scale. And it is the cosmic scale that we must use. It is not true that something is always immutably immoral. Most of immorality is a collection of socially acceptable trends whose outward explanations have no real bearing on reality or often even demand complete suspension of reality. Some are social guidelines learned from harsh experience and coerced into traditions that express safe advisories or direction. And some are the same Jiminy Cricket core that I have espoused, but cloaked in a ritual to enhance efficacy. ‘Thou shalt not commit adultery.’ Why? Because of the connotation of betrayal? Because an agreement of monogamy was violated, breaking a trust, depleting a loyalty?”

Melissa interjected, “Your case is, thou shalt not commit adultery because the limbic system in conjunction with the hindbrain can be stirred into counter survival behavior that diminishes the probability of the tribe's survival — the species' survival — and the least, a core-spondent's survival.”

“Exactly. In precise truth, the urge to have sex is a naturally occurring and irresistible force derived from survival enhancing mutation. It is, until recently, *the* essential motivation for survival. Sex with many partners is not counter-survival. Neither is its opposing force, the urge to restrict the activity of a prime mate in order to enhance opportunity for procreation — or more precisely, the pleasure of *doing it*. But, in an evolved social context, where there is no danger of extinction through having too small a gene pool, where technology optimizes for survival, social utilities have been invented using dogmatic edicts and rationalizations to diffuse the ripples of destruction that can occur if the limbic system is tweaked into over-zealous reaction. To accomplish this, we have attempted to impose memes of restriction. Adultery is a fabricated tradition of proscription. It is a meme whose purpose is to override a naturally produced hormone cocktail. And we have not learned to do this without calling upon some mystique. The power of sex has been too strong for simple reason to govern. We have even tried to employ the unseen All Powerful — and that has not worked either."

Melissa was clearly following the concept. "So, your Jiminy Cricket is designed to upgrade reasoning…it is Reason Version 2.0. And you want to use Mandelbrot Seed tech to get you there."

"Actually, I want to teach techniques to unenhanced members. But, I have also created a tool that makes it a lot easier to do the work for those soon to be enhanced. If you are in agreement, and we do begin selected enhancements, I would like the JCCI tested and made ready for distribution."

"How does it work? I confess to inventing no viable conjecture," she rarely uttered such words.

"I have developed a process of double blind testing of conceived action. With our modified brains, it is possible to partition quarantined self-acting models of the world, seed them with impulses or proposed behaviors and let them run on a high-speed timeline from identical starting conditions. Then sample them to determine which impulses/propositions best satisfy the Jiminy core objective. Most choices will be moot, having no meaningful survival impact. It will seem that these are disconnected from the special Conscience Engine. They are the choices with no measurable morality quotient. They will exhibit unrestricted variation. Some will fly in the face of tradition and social convention. But just because they do, does not justify their being classified as morality issues. The Cricket Engine will produce information in other cases. It will stimulate a sensation of conflict as sensory specific signals accompanied with an annotated decision tree, models, and example cases for consultation. The interface will adapt to the capabilities and style of the host. Choice will remain, but ignorance will disappear. The Engine will not choose or act."

"You will allow extinction, if it comes to that."

"Certainly. The possibility of extinction must be part of the solution set of the species. Put differently, it always *is* a member of the solution set. At their best, religious and social doctrines are designed to protect their children from the responsibility of survival…the complexities. But, an uncontrolled instinct of protection only increases vulnerability."

Melissa was compelled to agree. "Yes, there are some very twisted tools to enhance survival probability. Mutually Assured Destruction comes to mind. JCCE would reframe it to mean Mutually Assured Survival?"

"It would. There are worse ideas that have done much more damage than that. I know of a community that cut down the trees in a playground so children couldn't fall out of them. No record existed of such an accident at the time of the action. It was a preemptive act."

"I give you Mutually Assured Survival and you give me Preemptive Playground Tree Removal?"

"The same idea is behind jaywalking laws and preemptive military strikes. Protect people because they are not able to protect themselves. Put rubber baby buggy bumpers on every sharp object and every idea in the world. Consider traffic laws. It is only necessary to punish people for driving on the wrong side of the road, or for making unsafe lane changes because people do not understand, even remotely, the nature of driving an automobile. They know, more or less, how to control the vehicle. But they have not yet learned that driving is a team sport. And there is only one team. They do not know that driving and living are both games of Mutually Assured Survival. And the fellers of park trees, at the same time they severed the limbs of those 100-year-old trees, they were pruning the minds of the kids in that playground."

"It is inadequate training. And many experiments have shown that people whose decision making opportunity is limited, make less effective decisions." Melissa had done some quick research. "You are not intending to apply auto-morality with your little Jiminy; just advance the art of self-determination with some add-on tools."

"Again, precisely considered. Tech that could override any action that brings us close to extinction is one of those rare things that is not possible. We must come to terms with the fact that this isn't a task that can be foisted on miracle tech or dissociating belief systems. Extinction is an anchor at one end of the choice continuum. It is a spike representing death — nothingness, driven into the bedrock of existence. Extinction grasps one end of a tightrope leading away from it into the domain of survival…evolution and the future. Humanity is walking the rope pushing the anchor of the other end as they go."

"Hmm, I am reluctant to adopt such a meme. Still, it is most interesting, and better, by far, than any solution I have considered. Because of my enhancements, I have been blinded to the need for such an expansive solution. I do not find moral issues particularly challenging. But, now that you adjust the frame of the problem space, I see that it is the limitation of the normal human brain that is the real problem. It is, specifically, the inability of the brain to concurrently manage more than nine variables that dictates a solution able to chunk the more complex moral issues into reasoned pieces. Your adultery example is too easy. And as easy as it is, it is still a problem many are incapable of solving; but it is within a range of difficulty that could be considered 'trainable.' People could be taught skills enabling adultery to be a natural Cricket — solvable with a natural human brain."

"Yes, I admit to selecting a simple example for illustration." My extended years allow me to see that most moral problems have eluded definitive solution. Progress can be illustrated, but it has been slow, and limits of the brain itself are the cause. I expanded the scope just a little: "On top of everything else, the brain becomes less able as it ages. The ability to deal with distractions, degraded memory and entrenched learning itself further reduce the capability of the brain to manage complexity at full speed. These give way to behavior calcification — liberals become elitist closet bigots, conservatives become nostalgic rationalizing morality fascists. But, living less than a century has imposed a natural cleansing. Mortality is a natural limit cycle because, over time, flexible and energetic lives siphon away the power of the diminished minds in a predator/prey ecology — counter-evolutionary moral systems are eventually consumed. But the herky-jerky, feast/

famine balance is about to be perturbed too rapidly for the existing ecology — already popping the threads of its undergarments — to absorb. A new, unnatural resource is needed to get the species through the coming upsizing.

"All that being said, a facile personality combined with accumulated wisdom gleaned from simple longevity has a value. I do not refer to the flaccid rationalization of an establishment grasping synthetic artifices of logic to sustain, at any cost, structures supporting its garnered power — that is only mantric knowledge, not wisdom. That is an example of the organism being unaware of its own degraded performance. It is a confusion between the feeling of a well-understood process-experience and a well purposed process-of-reason. It is a chant cycling below levels of conscious volition. It is the song *It's a Small World* looping out of control because it was learned in a voyage of heightened concentration without awareness that the learning would outlive appropriate usage — over and over and over again. I am suggesting that certain capabilities of a well formed personality, coupled with broad experience may resist or postpone the decay of reason — at least, I am holding such an hypothesis."

Melissa gave a quick nod. She thought the distinction went without saying. "So, for this reason, you intend to upgrade the brain…because you have diagnosed the problem to be the organism itself." She understood. Before her enhancements she coped with morality like everyone else: with difficulty. Lately, however, she solved morality issues with the same genius she solved everything else. It was a very difficult problem, indeed, that was too complex for her. "I had forgotten what it was like to be challenged by things like adultery, honesty, gluttony, greed, racism, sexism, whatever 'ism.' I have missed the organic cause completely. Again, I have been affected by an emergent system. An out-of-awareness system is being moral for me."

"One man's saucer is another man's teacup," I observed. I don't know what that means, in this context, but it seems like the right thing to say.

"True, though of no particular import to the current topic. Returning to your project, let me say that I look forward to evaluating the seed when you are ready. When do you want to load the Mandelbrot utility?" She was happy to help me to my next step. There were things she wanted me to know that couldn't be communicated through our beads. Other MKS utilities would also forward my goal. I had work to do before I could turn this loose on anyone. To date, my experiments had all been within my own adjusted organism and my somewhat out-of-the-ordinary mind. I needed a tool that would help me work closer to the norm.

"Anytime. What kind of procedure does it entail?"

"It is quite painless. I have been protecting you from infection all this time. Trust me, I will make the experience a pleasure."

Just so.

Reporting In

CI_132 L0ARII:

In his office in New Jersey, the Agent read a cheery report from the operation Lead. Three operatives were enrolled in a Rivendell Intensive Business School course. A dummy corporation had been set up to front the operation. If, as was suspected, Melissa was in some way connected with the amazing new town that thrust up in the British Columbia wilderness, it was quite obviously prudent to avoid revealing the Agency's interest.

The team was actually quite well suited to participate in the course itself. Each was a college graduate with documented postgraduate work. They would not be detected as duplicitous through any deficiency regarding the demands of the course. The Lead was instructed to openly discuss the unusual Inclusionary Reporting policy imposed by the corporation, which allowed other members of the bogus company, not actually attending the course, to learn about the institution being attended. This was, ostensibly, to improve the opportunities for employees when taking advantage of the very progressive corporate sponsored continuing education program.

Sponsored Education Program Inclusionary Report
Topic: Project Management Styles vs. Objectives

Attendees: New Projects Development Team
Alfred Lundquist
Bob Montgomery
Cisily Nadero
Author: Alfred Lundquist

Rivendell Guides greeted us after we navigated Customs. From that moment the Guides managed a seamless succession of logistical elements that terminated with our luggage and ourselves in our rooms at the Lodge. No extraneous steps or cues were experienced. Refreshments were waiting in each room and an invitation to the "Class 'Meet Me' Social" was extended. Bob and Cisely attended and used the opportunity to gather histories of other attendees. One never knows whom one might meet...good opportunity for recruitment. I took the time to explore a little of the surroundings. What a marvelous place, and I really mean that! Later we met for dinner and a little wandering. It is a good thing we did; the course was very fast-track and ate up as much time as we cared to devote to it.

I was hoping to make this report earlier, and a little at a time, as I went through the course, but the truth is I became completely immersed in the material and have found it applies to many more corporate processes than I had considered. Also, there are technologies in use that I do not fully understand, but present very new ways of modeling objectives and management styles. I've had to reconsider nearly everything I have known about how to run a project. This made the Inclusionary Report particularly difficult because of the need to translate these new ideas into the language of old ideas. In the end, I decided there is no way to do this. I have gained a strong new model that will be very useful, but you will just have to come here to see it for yourself.

A word of warning: if you come here it will change your life!

Regards,

Al on behalf of Bob and Cisily

That was the net result of three masterful agents spending a very expensive week in Rivendell II. They found nothing, unless you consider the following statements for what they really are:

> From that moment the Guides managed a seamless succession of logistical elements that terminated with our luggage and ourselves in our rooms at the Lodge. No extraneous steps or cues were experienced.

Read: At no time was our equipment out of our view; the logistics afforded no surreptitious searches by us or the targets.

Refreshments were waiting in each room and an invitation to the "Class 'Meet Me' Social" was extended.

Read: We were not forced to participate in activities. If they wanted to search our luggage, they were patient.

Bob and Cisely attended and used the opportunity to gather histories of other attendees. One never knows whom one might meet…good opportunity for recruitment.

Read: We split tasks. Agent 2 and Agent 3 attended the gathering in case the Target might be in attendance and to gather intelligence on other participants. Relationships to be used during investigations were fostered for later use.

What a marvelous place, and I really mean that!

Read: Security is so good, I couldn't find it. If there are high-level security measures in use, they are beyond me and undetectable.

I took the time to explore a little of the surroundings.

Read: I swept the rooms for bugs and cameras and searched for security installations on nearby grounds.

Later we met for dinner and a little wandering. It is a good thing we did; the course was very fast-track and ate up as much time as we cared to devote to it.

Read: Later, we repeated the sweep as a team and attempted to locate evidence of high-security. We confirmed, as a group, my observations; we found no evidence.

I was hoping to make this report earlier, and a little at a time, as I went through the course, but the truth is I became completely immersed in the material.

Read: We were unable to report even the smallest infiltration success. We are either over our heads, or on the wrong track. I think you are right, we are surrounded by something, we are over our heads.

. . . it applies to many more corporate processes than I had considered.

Read: This is much more important and far reaching than imagined.

. . . there are technologies in use that I do not fully understand....

Read: There are technologies not of this planet.

. . . very new ways of modeling processes and management styles.

Read: Non-traditional thinking is in action by people with unforeseeable agendas.

I've had to reconsider nearly everything I have known....

Read: I have no idea what is going on here.

This made the Inclusionary Report particularly difficult because of the need to translate these new ideas into the language of old ideas.

Read: What little I have learned, you would not understand.

In the end, I decided there is no way to do this.

Read: I am not going to try to explain further.

. . . you will just have to come here to see it for yourself.

Read: We have failed to meet the objectives of the operation...you will have to do this one yourself.

And then, the last and ominous: "A word of warning: if you come here it will change your life!"

Read: Warning, you will be at risk if you do come here!

An accompanying e-mail contained a "Wish you were here!" jpg image of the three agents surrounded by "wild" animals while standing in front of the Lodge. Buried in the file was an stenographically encrypted communication. The Lead wrote a disturbing message.

Sir,

Bob, Cisely and I want to thank you for the opportunities you have placed before us over the years. Please now accept our resignations and remove us from the consultancy's Active Operative list.

We would like to express our regrets at failing this operation by offering to forgo payment of any fees or expense reimbursements accrued by this exercise.

Succinctly, the contacts made here are allowing us to test our abilities in a more secure career, namely marketing a revolutionary new product for a region of Europe. We, as you know, are well suited to function in Europe. While we are now convinced there is more to this school than meets the casual student, we are in agreement that it is, at the worst, benign. The opportunity is real and potent.

We can be reached via "Magic Color Sticks — Euro" should need arise.

With respect and gratitude,

A, B, C

P.S. The Inclusionary Report should be taken with strict literal interpretation. We had no meaningful intelligence to communicate. It was sent only because during the operation we had discussed it with others. I felt it best to maintain cover just in case you wished to continue advancing your objectives.

The Agent began to give careful consideration to the wisdom of continuing his investigation. Somehow Rivendell II had accomplished the unthinkable. It had made good on the statement, "If you come here it will change your life!" In a single week, it had changed the lives of hardened field agents. This was too close to a miracle to be ignored.

The agents had behaved honorably, attempting to communicate what they had learned, as little as it was. They had maintained their cover, leaving him with opportunities to follow them to the resort. He decided it would be better to abandon the cover organization and start fresh. If their ruse had been penetrated without their knowing, he would be naked. It would be prudent to use a different cover.

A Walk to a Meeting

There are times when internal perception varies from one's norm. As I walked from the Redoubt elevator towards the meeting, I was in a mood to ponder the wonder of Rivendell. Rivendell II was a facade. But it wasn't *just* a face. It was a weave of experiments, virtually all intentional. It was a testing ground. It was a meme infectress. It was the living proof of greener grass. It was a test bed for experiments in minimal bureaucracy. It was the probe, lodged deep, sampling the character of First Minds lured to it from the world's unconscious societies. It reminded me of many inspirational flashes of time I have experienced in my life. Those flashes winked out, sometimes leaving a legacy, sometimes just dimming out into the darkness. Rivendell was different from these. While it had begun as an improbable accident, it had transformed into a nearly self-sustaining engine of evolution. It might never produce more energy than it consumes...more likely it will become asymptotic, always needing a little push from entities out of the ordinary, never quite sustaining itself, always hungry for metaprobing. Without a meta-analyst, all successful innovation becomes ordinary. Meta-analysis is the equivalent of dusting fossils with a paintbrush; it reveals definition. If it is true that nature doesn't bother to evolve a species past its limited definition of success, that task is left to the meta-analyst. We had looked inward and attempted to outthink the limits of nature.

We were not arrogant, though the accusation would be made, sooner or later. We were being pushed by accidents of nature. Nature had jerry-rigged a souped-up forebrain to make food a little easier to gather, to enhance the chances of surviving through reproduction. Nature has no meta-analyst, it doesn't look forward. In all my years, I have never seen any evidence of nature saying, "I'd better keep things as they are, change might turn out badly." Our little nature is to use what we have. We are not arrogant to think, it is our little nature, our nature within nature. It was interesting to me when I realized that it was this power of our forebrain linked with our frightened hindbrain that produced most of the cycles of history. The forebrain achieved knowledge and change and the hindbrain became afraid and tried to stop it. If they had been separated, it would have been no contest, but connected as they are in a single organism, the hindbrain is smartened up considerably and can therefore be pretty clever in its efforts.

Main Entry: **in·qui·si·tion**
Pronunciation: "in-kw&-'zi-sh&n, "i[ng]-
Function: ***noun***
Etymology: Middle English ***inquisicioun,*** from Middle French ***inquisition,*** from Latin ***inquisition-, inquisitio,*** from ***inquirere***
Date: 14th century
1 : the act of inquiring : **EXAMINATION**
2 : a judicial or official *inquiry* or examination usually before a jury; *also*: the finding of the jury
3 a capitalized : a former Roman Catholic tribunal for the discovery and punishment of heresy **b :** an investigation conducted with little regard for individual rights **c :** a severe questioning
- **in·qui·si·tion·al** /-'zi-sh(&-)n&l/ ***adjective***

The Inquisition is the perfect example. Without convoluted rationalizations, the Inquisition would not have been nearly as exciting. Inquisitors made the word *arrogant* common language, referring to heretics...inquisitors themselves were shielded from arrogance by divine right. But the Inquisition would just have been a different form of barbarian horde without elements of sophistry, creative convolution and rationalization. A stupid brute, it would have attacked the ordinary and intellectual alike, and it would have left little imprint in history. It was the forebrain that was employed by the scared little hormone dumping, gut wrenching, heart thumping hindbrain that made the Inquisition what it was — a vicious attack on wondering itself. Since escaping the Inquisition, I have reclassified the Barbarians to having exhibited even-handedness.

We are not arrogant, we are driven by possibility. We are making cognitive effort to step around the tricks and traps placed on our path by our legacy organics. Our strides are long, because they can be. And in Rivendell II, with our ability to attend to detail, wonder itself is raised to a proper place among the achievements of our little natures.

An Odd Feeling...Was Felt

CI_132 L0ARII:

The Agent felt a twinge as he pushed open the door to the pub. It was the only occasion to tweak his sensitive nerves since he had been in Rivendell II. After a week of attending classes, roaming, questioning and observing, not a single stimulus said this place was anything more than an extraordinary and ambitious development by extremely clever business persons. The city made a profit. It paid for infrastructure and expansion from the enterprises that were permitted to participate. Each enterprise fit as a needed piece in a complex puzzle. Aside from the brilliance of its success there was no evidence the city/resort/school was anything other than what it appeared to be...that is, if you ignored the Frolicking Scholar Pub.

Rivendell II was a big and active place, so the Agent forgave himself for not noticing it before. Now, as he waited to depart, he had stumbled upon the pub purely by accident. He had needed a place to waste a little time before taking a shuttle to the airfield. He entered and tried to convince himself this was only his reaction to failure and error, a last gasp at uncovering a deception. He sat in a booth as far away from the men's room as he could get. The waiter appeared; the Agent ordered a plate of sausages and an interesting sounding local beer made somewhere in the Province.

"Excellent choice," the waiter said and departed the booth.

The Agent aimlessly flipped through the song tabs of the wall-box attached to the paneling of the booth. He marveled at the existence of Rivendell II. If anyone had asked him if such a place could be developed, he would have declared it impossible and not thought about it again. Yet, here it was, a happy place in the middle of nowhere. A happy place in a world gradually becoming more and more somber; a world being taught to fear by the consultants to media entrepreneurs of a different era than this jukebox. Or was he just shading his perceptions of the world with his own mood? The agent could not decide whether he was happy he had failed, or sad. It would require some time to tickle out the truth. But, he had to acknowledge the evidence: Melissa and the "Professor" simply were not to be found here.

Doors Open

Thodkin met Melissa just at the entrance to the Pub. It was possible for them to be in communication at anytime, so greetings were very subtle and silent. Thodkin pushed the heavy outer door open, permitting Melissa to enter first. The proprietor came out from the kitchen and gave them a personal greeting. Everyone in Rivendell knew the two, but none knew they were the city's founding parents. In this case, the proprietor owed Thodkin a special debt. It was Thodkin who had approached him in Princeton and convinced him to apply for acceptance into the Rivendell Business Consortium. He never did learn that Thodkin had already paved the way for him with the Consortium. Thodkin perceptively foresaw Rivendell's looming need for excellent potato salad and sausages.

After a moment of reassurance that they wouldn't need a table, the owner went back to the kitchen and began preparing a special feast for these favored patrons. To the wonderment of all waiters and waitresses the two were known to be the biggest eaters in the city. It was a matter of considerable conjecture amongst the various restaurateurs as to why the two never seemed to gain weight. Records were kept in mock secrecy regarding the meals and quantities consumed by Thodkin and Melissa. On a visit to Rivendell, Harley Travis had taken a quick statistical sample of their consumption quantities per restaurant and had one of the Redoubt workshops fabricate a large granite empty bowl with a ladle-sized spoon abandoned in it. It was a work of art perched on a short but elegant pedestal. One day, with great hoopla, it was awarded to the alleged winner, the restaurant that contributed the most calories to the pair. Since then, the award had been passed amongst several restaurants and was displayed prominently and with a cheerfully puff-chested pride. The bragging rights were worth a small additional clientèle from locals seeking to discover secrets and culinary specialities, but it was really all in fun. Still, it was true that the restaurants worked secretly to keep Melissa and Thodkin entertained with new quality dishes. The special attention was given even though the two had never been anything but complimentary in public. Interested persons kept a close eye for signs of cuisine burnout and employed various tricks to lure their targets in to sample each new delicacy.

Thodkin led the way to the booth now inhabited by the Agent. The plate of steaming sausages, was, as yet untouched, and a tall chilled beer glass, still sported a good-sized head. The Agent looked up from his idle searching of songs.

"Permit me to introduce Melissa...and my name is Thodkin. I believe you have been looking for us?" Thodkin waited as the Agent gathered himself. Many seconds passed as he struggled to take the first steps in a long journey. "Would it be too intrusive if we joined you for a meal? Perhaps you have questions we could answer while enjoying the Pub's special fare?" They accepted his silence as permission and slid into the booth facing him.

At last the Agent managed to speak. Looking at Melissa he said, "You are younger than I imagined."

"Yes, you probably ran aging algorithms on college pictures? You began seeking me about six years ago?" Melissa knew exactly how many years and days ago the Agent had begun his search, but had taken to placing slight imprecisions in her dialog to soften the contrast between her communications and the rest of the population.

"Yes, nearly seven now, but it is really something in your face that surprises me. I expected to see...well, I guess a geeky naïveté to give in to vernacular...clearly I have failed to take some forces into account. And, if I am not overstepping bounds, your physicality seems nearly athletic, which is not as I imagined."

"Much has transpired since Princeton," Melissa agreed.

The Agent looked at his sausage, it was just a delaying tactic while he gathered his thoughts. Thodkin helped him out.

"Please, don't let the bangers get too cold. As you probably know, they are derived from the style made by butchers in the Lincolnshire wolds. They are rather famous in England. Our own food is on its way, and feel free to sample what arrives. There will exist some special treats, to be sure."

The Agent took the opportunity to slice a bite and somewhat tentatively pierce it with his fork. A foreboding possibility had just occurred to him. Again Thodkin came to his rescue. "While it is true we have known of your presence and some other details, you are in no danger. Eat the food and enjoy it. Perhaps you will wish to remark about the fine quality of the sausages at Rivendell when you return home."

At that very moment, the Agent had been wondering if he would ever see his home again. It had filled him with a profound and mournful feeling. The strength of the emotion was surprising to him. He had been on many many field operations and could not remember ever feeling so exposed as he did at this moment. Never had his target turned the tables on him. Never had any adversary matched his attention to detail, his ability to plan, to provide for contingencies. Now a college student and a — what, certainly not a vagrant — had caught him trousers-at-the-ankle; more brutally, completely naked.

The agent rallied slightly. "On the plus side, however I have misjudged the risks, it is nice to know I was not in as great error as I had come to accept."

"Yes, the bright side," said Thodkin. "However, the implication that you are at risk is an error you are still sustaining. You are not at risk, nor have you ever been at risk. You have never threatened us. And even if you had, we would never need to harm you, nor choose to. This is not a plot created for entertainment nor a sensationalized media report designed to bolster ratings and sell advertising."

As comforting as this was in one way, it was disconcerting in another. It meant that he and his family were safe, but also that he was not as formidable as he believed. They had no concern about him harming their enterprises, much less any fear of him.

Several waiters arrived, providing a useful moment of strategic thinking for those who might need it. The table was soon filled to capacity with food and drink. The chef and owner arrived to advise Melissa that she was to try this dish and that, paying attention to combining it with such and such a beer. The advice puzzled the Agent. It went on, dish after dish, surely they didn't expect her to eat all this food. Perhaps she could sample each dish, but it would be like having a meal of appetizers. In response to the earnest culinary education, Melissa sent forth a smile that blazed like an early nova and had such irresistible gravity that the hearts of the Agent, chef and the attendants were swept into it. She thanked her advisers and assured them she would follow their instructions to the letter. Satisfied, they withdrew to the kitchen after advising the bartender which beers to provide and in what order. Nothing was taken for granted.

As if compulsed, Thodkin and Melissa began to eat. After a few gulping, full cheeked bites, oddly taken in exact accordance to the chef's instructions, and a full draft of beer, Melissa asked the Agent: "What is it that kept you interested in us for so long, given only the vague and unsubstantiated evidence you garnered from your clever programs."

Again the Agent heard the message that he was overmatched. They knew precise details about him and he had only vague patterns and statistical innuendo to guide him. What should he tell them? Did he have any advantages? Did he have any cards to play to try to recover some semblance of parity, some quality of being a peer? He was considered a master in espionage. But he felt unranked.

Thodkin made yet another effort to ease the Agent's cycle towards fear. "Please, before you answer, take a moment to fully understand. We are here to learn from you. Only one other person, not of our explicit choosing, is aware that Melissa is in any way responsible for what you have eked out in your investigation."

"You refer to the ex-President of Princeton? I am aware she suffers something akin to post traumatic shock from what she experienced the day in the lab at Princeton."

"Yes, she is the one. We keep watch over her recovery, and have pulled a few strings to ensure she has the best care and that her stories don't result in a misdiagnosis of her condition. After all, what she is traumatized by was a real occurrence that will not be considered reality by unaided practitioners trying to help her."

"If she were believed would it not threaten you?" The Agent was a little skeptical that Melissa and Thodkin had the President's well being in mind.

"Sooner or later, she will be vindicated. But our plans proceed in precise measures, so vindication will not be immediately forthcoming. In the meantime, we make sure that she is not asked to give up what she knows to be true. She is being coached only on how to improve her ability to adapt to her circumstances. To help her prevent her fears from driving her actions. And to shed some of the cynicism that took away her purity of purpose."

It was with a measure of wonder that the Agent watched Melissa eat as Thodkin was speaking. She cleaned dish after dish and always conforming exactly to the instructions of the chef. The waiters arrived with precision at the exact moment a new beverage was required, taking away the empty glass from the previous course. Melissa must have wolfed down three meals in the short time Thodkin spoke. Then she slowed down, allowing Thodkin to eat while she redirected the conversation.

"If I may ask again, and please tell me if I form the query imprecisely, what has kept you on our trail for so long? And you may skip admitting hope for some sort of fiscal benefit; it goes without saying that you must seek financial rewards from your effort."

The Agent finally came to a decision. He could find no downside to explaining himself as clearly or bluntly as he wished. "At first, it was simply that I had failed the President." He gathered himself and continued. "Our investigations indicated that something had taken place that was not, to understate it, ordinary.

"The college security cops, normally adequate witnesses, were at once frightened and confused. They produced a convoluted description of what transpired when they had routinely checked the labs for being locked. They often find the labs unlocked, and simply lock them and move along. Being a security cop is most often unexciting. In a university it can be a little more fun, but rounds are usually boring, except for the occasional interrupting of a tryst in progress. An unlocked door does not usually lead to pink-faced students. So, these two guys were not prepared for any kind of action, much less being directed by the University's President to arrest anyone. And when almost immediately a man dies, and a green animal is apparently attacking the University President, who is screaming at the top of her lungs for them to shoot it, they are pushed past their training and abilities. It is most fortunate that they shot nothing or it would have changed circumstances most publicly. I was able to filter a collaborated scenario from conversations with security that correlated with the President's self-serving account. I did not confront her with the discrepancies and at first treated it as a case of the President's wishing to take advantage of a brilliant student and forward some funding objective or the like. Of course, your amazing disappearance soon changed my mind."

"How did you explain the Green Hair Comber?" Melissa asked.

"Is that what it was? We never learned anything about it. I had to satisfy myself by accepting that if it really did exist, it escaped, or disintegrated completely during the altercation. The most likely possibility for escape was that you and Thodkin had taken it with you, although the witnesses were consistent in saying that you left while the thing was attached to the President's leg. Further, the windows of that lab do not open. All ventilation is accomplished through chemical and biofilters in case of toxic accidents. There was no evidence of a green monkey in the filters. Nothing. I rarely propose mass hallucination as objective analysis, so I relegated the green entity to a category of missing, misidentified or unknown species of animal and spent little time considering it. Should I have spent more effort on it?"

"No, you were correct. There was no way to correctly interpret the information you had," Melissa said after finishing a long swallow of beer. "Please continue your explanation if you don't mind disclosing your thoughts to us."

"I was able to track you to New York. I admit I must have made several errors in the process. After you left the train station, I believe you took a taxi. It is possible you spent a night at the Waldorf Astoria. Today, I believe you did, but at the time I never conceived you would be able to do so without leaving some obvious evidence to confirm it. I uncovered no such evidence and when I found your bags still in the Airport Lost and Found, I ceased tracking you with operatives. Your trail was cold, to coin a phrase. You had escaped my team; other means would be required. I found your disappearance confounding. Permit me to speak plainly; you did not seem up to accomplishing such a vanishing. Or, you were not who you seemed. I soon proved you were who you seemed, leaving only two possible conclusions...you had help or were vanished by someone." His glance at Thodkin communicated his difficulty in considering Thodkin as capable of such success.

Melissa did not take offense; she knew what he said was true. "You are quite correct. I was very nearly totally stupefied during those early days. Perhaps we will talk of that sometime later." She continued eating, though the table was beginning to look like a camp of lumberjacks had just gotten up from lunch.

"I don't have much more to tell. I have experience with what it takes to disappear. It is difficult in so many ways that those living are nearly always found. Yet, you were not to be found. How is this possible? In the end, and to finally answer your question, I simply had to have an answer. It became a hobby. And it was a useful one. I learned many new skills simply because I had exhausted all my others and still not found you."

Thodkin slipped in an unrevealed piece. "Your interest was also sustained by the work coming from the other 'missing students'?"

Again the Agent felt the sense of fragile nakedness. "Yes. The business community knows the ToolShops (your ToolShops) are interrelated. But, they are without the knowledge that each has key founding members from the lost students. Only the coincidental, mass dropout sustained me. That and the phenomenal successes that were forthcoming from them."

"And the President?" Thodkin prompted.

"You are right. I prefer not to speak about clients, but I can say this; it takes a special person to become a university president. I knew her well, though not personally. She was a good customer with a clever political mind. She often utilized my services to turn a disastrous student goof-up into a profitable gratitude for the University. Something unprecedented, forgive the pun, had happened. She was extraordinary, therefore the event was more so. I reminded myself of this fact many times over the years to keep from becoming lackadaisical about my hobby.

"But, Rivendell was the best evidence, or so I became convinced. The ToolShops were everywhere in Rivendell. If the missing students were connected to Melissa and they were so key to Rivendell, I concluded that if Melissa were to be found, it would be here. And if I had been chasing nothing all these years, I would find that out here. In the last couple of years I have begun to doubt myself. The world is not a very nice place. It deals rather swiftly with the extraordinary. You see it everywhere. Scientists are not immune. Listen to any panel of leading scientist discuss cloning of anything human. I have interviews in my archives of scientists asked what the species will be like 1,000 years from now; what will humans be like in 1,000 years? They, without exception, dissemble. They hide the picture of their expectations, painting it over with falsely incremental changes of the human organism. I am not a scientist — well perhaps I might be considered a computer scientist — but I have educated myself in bioengineering with the hope of discovering your...Green Hair Comber in some dark corner of research, and one thing is clear to me: if our species survives another 100 years, much less 1,000 years, it will be nothing like what we know it to be. Without exception, these scientists cover their collective asses — which is to say, funding — by sparing the world from trying to imagine its impending future."

Melissa asked him the hard question. "So you came to Rivendell thinking we hid the future...why; to keep for ourselves? To exploit? To avoid political retribution or conservative ire? Perhaps we fear the witch hunt?"

"Forgive me if I don't believe in the beneficence of others."

"Ahhh," said Thodkin. "So much said in such a small sentence." He added a touch of pretension to the comment. "But first, thank you for speaking with us so frankly. While we have more data than you could imagine about your nearly seven-year quest, only you could provide us with your thinking, insights and motivations. Now, perhaps you would allow me to speak as frankly?"

"Certainly." The Agent waited. He was relieved to be able to do some listening.

"To start, there are three imprecise, subjective words in your earlier sentence. Forgive, believe and beneficence. They are a reflecting mask lacquered with the faux substance of hard fought wisdom being used to shroud a well constructed cynicism. It is a discussion about the possibilities facing humanity and the entropic counter force, cynicism, that has motivated us to share lunch with you."

Perhaps, It Is Time for a Story

"Perhaps you will indulge me by allowing me to speak circuitously for a moment or two. It will be counter to your preference for more linear forms of discussion, but I believe it will be revealing for you." Thodkin asked permission to tell the story in his own unusual way.

"My flight is not for some time yet. Please do as you wish." The Agent prepared his mind for a logical onslaught. He was expert at piercing intellectual strategies of persuasion. He took a quick survey of his situation:

It was clear that Melissa and Thodkin were not here because of the food, although they seemed to enjoy it and were treated well. His own body did not seem to be poisoned or drugged. He assessed his faculties and could find no evidence that his defenses were impeached. He was sporting a slight adrenaline elevation from the challenges of the moment and he felt alert, even invigorated. He was doing very well considering all the setbacks he had suffered and the uncertainties that remained unspoken.

Melissa, for no apparent reason, reached out a comforting hand and gently patted the Agent's wrist three times and withdrew it. The Agent simply looked probingly for something he couldn't find in her eyes. The gesture was for all appearances an expression of support or caring or perhaps understanding. Her hand was noticeably hot and left a lasting sensation where it had touched him. He masked a moment of discomfiture with a sip from his glass of beer, still his first. He noted that both Thodkin and Melissa should have been severely impaired by the beer they had guzzled with their meal — meals — but he saw no outward signs of intoxication.

The Story

I was first told this story by a person who may or may not have been subject to forces similar to those in your work. He was told the story by his father who had first heard it at the knee of a village elder, who might have been his grandfather, and he told it in just this way to me in another time.

It is a story of a tested and true strategist and versatile tactician who plied his skills to help a Feudal Lord, a true Land Baron. The planner was the golden key to maintaining the quality of life the Baron had become accustomed to and knew to be his right in the ill-considered model of the cosmos held by royalty. Much has changed in the world since those times, but in other ways, little has…the Baron could be living today under the pseudonym,

CEO.

Each day the Strategist confronted intrigue, mystery and disguised goals dressed up in costumes of cheerful deception.

An educated man, which for the times meant well-read, he studied the ideas of the world. He received exotic books as gifts from friends or those seeking their own advantage through sociable bribery. The books were from scholars, dilettantes, religionists and crackpots living in distant cultures and each held foreign concepts…the authors and the books. The ideas he found in the books he considered thoroughly, sometimes rereading a work a hundred times to extract meaning, or sometimes studying and searching his library or the vast unread libraries of the Baron for other thoughts in order to discover a counter argument to contentious propositions from an outworlder's mind.

Involved in his studies and his work, he still had moments of introspection in which he looked, as if from outside himself, and applied his skills of perception and analysis to his own motivations and behaviors. He explored the demonstrations of his life formulations as their effects were exhibited in the real world of the senses.

On one of these days of introspection, the Strategist took a walk up a long path to the top of a hill that would give him a view of the city now fully under the Baron's control. A city that the Strategist had caused to exist.

As he walked the path he was gradually released from the efforts of the day. Gently a calm golden shadow-self made its way out of him. It had awaited this rare opportunity, and coalesced as a nearly visible manifestation, easily keeping pace with the steady travel. It seemed to the Strategist that he, himself, was floating there, removed from the visceral motivations of the body, able to weigh the minutiae of living, without bias, without pushes and pulls from need and desire.

Thodkin paused for a moment to stack an empty dish on another, clearing a little space for his elbows. His head tilted down a moment, then he gently looked up into the eyes of the Agent and continued in a matter-of-fact tone as if conversation was managed by stories like this every day. For his part, the Agent showed only interest in the story, patiently waiting for the attack he knew would come, for what purpose he had only guesses. But, in opposition to his own self-expectation, as he waited he shifted away from any apprehension, instead tending toward calm where his breathing was able to become regular and gentle. He felt a profound relaxation that rose to a state of clarity and focus of such acuity that it pushed out all sensation of the room and indeed spurious thought. Only the story was allowed attention.

Thodkin resumed his narrative.

The Strategist had learned the motivations of each and every important or influential member of the city; such was his capacity. He had categorized and cataloged foibles and fetishes, dreams and dementias, aversions and avarices. It was a consuming vocation and he was good at it.

And that day, as he walked, the path meandered as it always did, along the contours of the hill. It is true of most paths that they take the way that is easiest. So, much of the path showed no horizon, only the earth walls made from the cutting of the way. Only the man-made ruts were in view for much of his traveling. But, when the turns worked just so, the Strategist was able to see bits of the city. At first, he saw just the tallest spires. They gave him a feeling of promise and a sense of wonder at the ability of humanity to accomplish difficult feats.

Farther and farther up the hill he walked. As he did so, the twists and turns that on occasion revealed a vista, showed him more and more of the city, until at last he came to the promontory he was seeking and walked straight to its border. From there he looked down at the city. But, the sight no longer gave him a sense of promise. Now he could see the curling smoke of pollutants and streets of dark shadow even in the daytime. He saw becloaked dwellers scurrying, hunched down, uncertain of their safety while at the same time heads were swinging from side to side, seeking any advantage over others. From this place on the hill the darkened regions seemed to reach up and dust the spires of achievement with gray soot. As he looked down, he saw the city as he knew it was, as he expected it to be.

"Nice place to live?" The Strategist was startled to find his attention had been so constrained to his inner landscape that he had not noticed the promontory was already occupied. A young man, dressed in baggy pants of the time and shielded against the brisk air with a plain but utilitarian cloak sat facing towards the city. A polished staff fashioned from shards of many different woods tightly interlocking to form an artful smooth branch lay beside him. There was little doubt the art belied the usefulness of the stick during travels. Near it an aged leather backpack lay on the ground, its lid stretched flat acting the part of an oft used picnic cloth for the cheese, fruit and goatskin canteen the young man had been attending. "You are from the city? Care for a bite of cheese and apple?" the young man inquired.

The Strategist recovered quickly from the disappointment in finding his quietude interrupted. Using an exaggerated full-lunged intake of air to communicate a profound enjoyment of the solitude, serenity and freshness of the hill, he gave a quick twitch of his head and a small open-handed flick of the wrist to indicate "No thanks, I'm just out for a walk, enjoying a moment away from it all in the beauty of this little-used spot." Without actually ever speaking, he answered the last question first and left the first two questions unanswered. He didn't really want conversation, and certainly didn't want to talk about the city. But, something about the young man caused him to relent. "No thank you, to your kind offer, and yes, I am from the city." That should be just enough, he thought, to curb further intercourse and he returned to his thoughts, keeping only a tiny attention on the stranger to ensure immunity from surprise attack.

The two men looked out over the city. Their thoughts circulated the foyer of their cultivated minds, shaking hands with one idea or another and moving along, waiting for the emergence of a profundity, insight or a visiting epiphany that could press open the heavy doors to the sanctuary. For both men it was this way. Thinking was a religious experience. The ability to think was held in nearly egoless reverence and the purification of its process was a deeply anchored commitment at the end of a muscular chain; but as was to be expected, there were differences.

"Please, forgive me for intruding…I can see that I am…but I haven't been this way for many years and, if you will permit me to observe, from superficial evidence it seems possible you could advise me on a minor detail. I am thinking of stopping in the city to seek employment. Is the city as vital as it seems?"

To his own surprise, the Strategist quietly, but with vehemence, blurted, "It is an organ of festering pustules, of illiterate self-interested cancers." In a moment the golden calm accompaniment that had surrounded him quivered and shrank.

"So, it suffers as other cities of my travels," the young man observed.

"All cities of man share the sickness." The Strategist was succinct, but allowed a little of his disgruntlement — at finding his quiet place inhabited by an initiate — to edge the declaration. He prepared for a deviated response.

"Did you not participate?" The young man inquired. The delivery of this single direct question altered the form of the dialog. The Strategist had a finely tuned sensorium. He routinely opened closeted thoughts, wrested hidden agendas into the open, unmasked the disguises of intrigue and picked locks guarding treasured coins of varied denomination and species; but there was a deep well of innocent power fueling the traveler's voice. The clarity of that voice obscured even the Strategist's discernment of ill-shaped purpose. He could find no trace of malignancy. And quite outside of his intention, he answered without his shroud of caution.

"I once thought I caused the ulcerations."

"But, you did not."

"I have studied to determine the answer." The Strategist now spoke in a surreal waking dream. Somehow the young man had become the manifestation of his metaself — in the flesh. The young man had become the voice of the Strategist's rigorous, objective and farsighted mentalese. The voice was clear and impelling, but gentle and experienced. It seemed a voice beyond possibility for one so young. He heard the past and present and felt the voices merge into his own, forming a younger tone but infused with a correlated wisdom. The dream allowed a safe place for direct communication.

"You were rigorous."

"Yes."

"To the conclusion which you now live by."

"Yes. After an uncharacteristic struggle between desire and reality, I concluded that the human world is doomed by its nature. Monothematic self-interest (greed) and innate aggression will bring extinction…eventually; the time frame eludes me, but it is inevitable. It became clear to me that these roots curl in a death grip around the bowels of human nature; ever they tighten. Our nature is powering a relentless and extreme pressure that squeezes stinking waste products to the surface of all endeavors." The golden aura of the Strategist's monitor-self dimmed where it hovered nearby.

"Graphic and emphatic, although a well-fertilized error." The young man stated with a calm of profound golden rays, glowing softly from a sun of truth that had silently overtaken the other forces holding the promontory. Everywhere colors enriched, sounds achieved new melodic, and each thing reached toward the Strategist either in strong comforting gesture, or to flick away pesky diminished aspects perched on his demeanor.

The Strategist was aware the young man had just called his conclusions dung. "Where is the error?"

"Simply, aggression is kinetic energy. It is not intelligence, it is a component of ecology.

"Moreover, aggression begins as a giant stone sphere poised on a tall jutting mountain peak. In every direction below are possibilities for the sphere…strong attractions. Each hill and valley, each rolling sequence of mountains is a possibility until one force or another unbalances the aggression, when it will roll along the path of least resistance until the kinetic energy is equalized through the graspings of simple friction, the counter forces of that which it confronts and the force of gravity proportional to the stone's massive obesity. Good or bad, it has no will. It does not choose. Nature does not choose. A natural force cannot be cited as responsible for outcomes chosen by intelligences."

The Strategist had hoped that nature was to blame for his own descent into a fatigue of spirit. Surely, his intelligence was natural and wasn't it then a weakness of natural forces that brought him low? He made a quick venture through the library of his learning hoping to find a citation of reason for use to dispute the counter assertion that intelligence must outmaneuver the very nature it was an element of. The part must become greater than the sum. In the end he could only postpone acceptance by employing the tactic of the "sincere query." "By what means do you know this?".

"The knowledge is present in the definition of evolutionary intelligence. Intelligence is the ability of mind to choose how to employ nature beyond immediate primal survival need. This ability implies expertise. Without expertise, intelligence is nullified, plateauing and returning to elemental nature...instinct."

"Your giant stone is loose in the city. It cannot be stopped. It rolled down from an infinite pinnacle and carries the inertia of all history, banging from house to house squashing flat constructive reason. And greed is Atlas flicking the stone about like a child's marble, preventing the stone's rest." Again tendrils of void pushed their way out of a nearly exhausted piece of mind.

"History is too short to justify a conclusive trend analysis. And your greedy Atlas is Unreason who will loose all his marbles as skills of cognition are disseminated. The clacking marbles of unreason will bang about dissipating their energy until, in the end, they are exhausted of it." The traveler offered a matter of fact and the void lost cohesiveness.

"History is consistent," came the dark's reply. It seemed to the Strategist that light played back and forth over the little patch of earth where he and the traveler stood. From the traveler flooded golden hope, shaped to form by prudent, lucid thoughts and purpose. Then just as it neared him, the Strategist felt an unbidden parry lash up from a darkness to deflect the full comprehension of the offering. The dark place understood that it was important not to understand too well. Then just as the fragile dark was reestablished, the traveler would strike it again with illuminating truth, or hope or — worse — more reason. In the dark, fear rattled.

"History is inconsistent, distorted and in the main both recorded and analyzed without expertise. The organized recording of history is less than 10,000 years old. History is yet to become a science."

"Science?"

"Yes, a form of analysis in which a hypothesis is tested by experiments producing the yield of verifiable data."

"No wonder I have not heard the word. The practice does not exist in the real world as I know it. The city is clearly ruled by mysticism, charlatans, prestidigitation and slight of mind."

"And yet you are still a seeker," the traveler prompted.

"Now barely so. I am relegated to a sickbed where I attempt to immunize myself against advanced stages of moral infection. My treatments are failing, I am surprised to find out. I am developing resistance to the countermeasures as this desperate conversation with a stranger demonstrates. I ponder quarantine, but do not know its sterile location." Though he said these words, the Strategist found he was not fully present behind them. They seemed to stand off from himself in a translucent caricature. The image was aged and bent, but it was not substantial. In fact, he did not now feel the words as weighty truth, but uttered them anyway. The words whispered, reduced to simple autonomous verbal habit.

"You are orphaned from your kind. Why do you not leave the city?"

"Three reasons. I have profited by it, allowing me to provide well for my family. Leaving brings uncertainty in a world where sagacity and pursuit of counter-myopic force are a combination undervalued. Second, it is familiar. I have accepted it and learned tactics to cope with it. Third, I have researched documentation from cultures far and wide. In each I have discovered a consistency of two facts. The non illusory documentation depicts much the same reality as I suffer, and only works of fiction describe improved alternatives."

"You find the fiction still depicts hope?"

"Imagined and without reality."

"Imagination is the source of any possible species-driven elevation. You noticed imagination in reality. You identify the tenacious, perfectly faceted diamond of hope and have not harvested it."

"The diamond is not a carrot, and cannot, in any event, be utilized in solo."

"Attend your own data. Each culture you study also exhibits the fiction, yet you give weight only to deficiencies."

"Thoughts unaccompanied by action are impotent."

"So true."

Again darkness heard its best-kept fear shudder in confinement. "In the end, I am too fatigued by bruising daily existence to overcome my own impotence. At night I lie in a passionless bed hoping only to rest well enough to continue again the next day. My studies are fueled by habit rather than my once fevered belief that answers could be found if I scoured the work of other minds outside of the local epidemic. Now, I am more motivated to discover a universal constant as justification for my failure than to be confronted by an idea that might imply an action I could take to shift away from the path I tread with lemmings. I am compelled to admit this to a stranger for no known reason."

"Many answers exist. Most are hidden by the ecology holding dominance in your mind at present." The stranger spoke simple calm reason that did not allow wallowing in the confusion of symptomatology.

"It is a dry and dead place."

"You exist in an arid desert where storms swelling from colliding eddies of hidden premises have blown dunes of parched sand over the edifices of inventive wisdom. But, it is just another ecology, existing for no particular intendment. And it is not dead. It is inhabited by such creatures as are able to adapt to nature or are adapted by nature. One of these is you. You survive in this desert because you must care for your family and satisfy the tax collector who is your patron. How many others do you know that survive in it?"

"I know no others."

"You know your family. How many others."

"Certainly, fewer than a dozen."

"Then a statistical miscalculation is also at work in your desert. Besides the few you acknowledge, were there actually truth in the analogy of 'lemmings,' then among the lemmings are others. There are no lemmings; there are only others with fewer skills. Lemmings use two instinctive survival tools, herding, unfortunately coupled with an optimized ability to see minutiae in dirt. Circumstance may extend the attributes beyond the value point. They lack discrimination at a distance and trust their feelings of security when surrounded by their own species. Suboptimal ecologies of thought have insinuated minds, creating topologies that work to limit choice. Still, such minds, using sparse resources, are making the best choices that are available to them."

"They could choose differently."

"It is an error to think that any possibility can be spawned within any ecology of mind. Ecologies are not mythoi. You view others from heights of mind they may never reach but still you scuttle over the dunes with them, arms linked in a head-bobbing, shoulder-swaggering two-step motivated by engines of purpose that are nearly zygotic. The evidence of this existence — your footprints in the sand and those of a thousand dance partners — are soon blown too smooth for recognition in memory. Individuality is tumbled into history's mass grave of the anonymous, marked by the single stone that reads 'Who came before.' To alter the inevitability of this forecast, ecology must be changed.

"It is time to forgive the other dancers for exhibitions of limited choreography. It is time for you to invent archeology and patiently dig away the accumulated nonsense being used to oppress both beauty and happiness. It is time *now* to *realize* that intelligence devoid of action is impotent and to act upon the knowledge. And it is time *now* to understand the statistics of change. Change, evolution...the future...is created, good or bad, by choices."

Though speaking of an arid landscape, the voice of the traveler conveyed a counterpoint of lush and colorful environment overflowing with well-nourished flora, bathed in the sparkling light of sustainable utility, science, invention and art. Somehow, stories of achievement against all odds — classic stories of dedication and persistence — grew up out of the sand. Stories of everyday solution, humor and beauty reached out from the parched earth and transformed it, creating expansive vistas in any direction. Arid was fecund, dust was moisture, barren was fertility and impossible became just another task.

"My choices seem small and insignificant in comparison to the rabid carrier infecting everything I know." The Strategist had just made a weak admission of choice, of possibility. It was finally the case that the dark was able to enlighten a single white flower of hope.

"Each little choice enhances or restricts possibility. So rarely is a choice null as the occurrence to be statistically insignificant. No thought is zero; every thought counts."

"Please specify a little choice possible within my desert. I am not yet totally submissive. I can marshal energies to fuel at least one small evoluting behavior." From deep within him, where goodness protected his endangered forces of goodwill, desire for contribution and manifestations of value, a wide fan of patient roots nourished possibilities that had laid dormant. Possibility split apart the dark. Fresh air breathed into the center of the man and after many years of suppression the Strategist inhaled clean, justifiable hope.

"May I offer this? Just begin with language. Reverse the trend to exaggerate descriptive deficiencies of yourself and others. How do the inaccuracies help? Substitute curative language. Append the temporal qualifier to thoughts and observation."

"In specific?"

"Imagine coming upon a tall wall of slippery stone stretching across your path and out of sight. Going around the barrier is obviously futile, and your first attempt to scale the wall fails. A passer-by stops and inquires 'What is your status?' You might say 'I can't find a way over the wall.' The statement is constraining. Instead use, 'I haven't found a way over the wall, yet.' This statement permits and invites choice."

"It is a small thing, as you say. I believe the idea has other applications?"

"Yes, many. Imagine a confrontation. An example is presented of an affiliate's failure, in the extreme, to achieve an agreed-upon objective. During a discussion with those you serve you hear this quote, 'The man is an idiot!' I believe you may have experienced a similar enactment?"

"It has occurred."

"Select the following option when it is your turn to comment. 'In this instance, the man behaved the idiot.' An instance of behavior is a thing. Named behaviors may be considered to have borders defined by a span of time, most quite short; rarely is one of extended duration. However, the man is always process. He is a process defined by the collective namable and unnamed behaviors occupying the span of his life. Defining the man to be an instance of his behavior results in stagnation and discards other potential. It is more useful to acknowledge demonstrated alternatives found in the processes of living."

"Again, it is a small thing, as you say, yet employed pan-opportunity I sense it is profound. What brought you to value it?"

"Linguistic morphology is the first tool of change. I often must learn the new language of an idea before it can be expressed in my behavior. Language is either the enabling artisan of vague impressions of possibility, or it is the cageman fabricating ever stronger prisons around choice. Pass this on to your children. Show them the technique in your expressions. Pass it along to those you are permitted to counsel. It, as you say, is at once little and profound."

Was this all that was needed, thought the Strategist? He could not explain the unexpected patterns of thought he was experiencing, the new questioning and the sudden sense of an impediment removed. It appeared as truth that he could know regeneration.

The Strategist turned to look over the city. It seemed to him that amidst the shroud of soot he could make out his apartments. Just there the soot was pushed away by a gentle but insistent golden glow. As the soot retreated from his apartments, an attraction exerted against the diffuse gray form. It was invented from the rooted language of possibility and solution sourced by his potent, unleashed, creative imagination. The soot pulled toward its own. The layer of gray attached to buildings and natural spaces lifted up, gathering, coalescing in a shape of fog that moved away from the city. As it passed the outskirts, the less populated perimeters where the nearest farms and ranches spread away to hills and horizon, the soot began to transform. It began to change away from the gray. It absorbed the greens and yellows and oranges of the land, and the various blues of the sky, until it became another shape and sense altogether. He could watch as this new enriched form dwindled away into the distance until it was only a tiny glittering shape disappearing over the farthest horizon.

The Strategist breathed as he noticed the unexpected transformation. He felt calm as he felt the sense of horizon and the gentle sounds of living things. And still his imagination provided one more shift for his noticing. His attention was once again attracted to the place where the dot had receded from sight. A new aspect was returning. The colors and shape filled the Strategist with a remembered sense of comfort, safety and peace that gently expanded wide and then settled as a blanket over the city.

He felt the glow physically whenever his gaze swept over his neighborhood. It was an unlikely but palpable sensation. Perhaps it was his little fiction, but a secondary awareness of various other such spots of lesser intensity could also be discerned throughout the city. The Strategist just permitted his critical nature to rest in the cradle of these glowing, warm emissions. For a moment he rested. For a moment he sensed something of his youthful vigor and sense of wonder. It was right enough for now.

After a time, he breathed deeply, straightened into a more vigorous posture and spoke. "Perhaps, I could find some work for you in the city. I am not without influence there," the Strategist offered. But as he turned toward the young man, where he had sat was just a flattened patch of grass. In the distance could be seen only his cloak's hood and the top of a staff receding in a fluid and rhythmic motion along a path away from the city.

Pyramids

Thodkin paused; the story was concluded. Melissa observed the Agent. The style of the communication was so surprisingly indirect that Melissa marveled how the Agent had become so transfixed by it. She decided this spoke well of him. The Agent sat head down in thought considering possibilities, finally becoming the first to speak. He looked up.

"A story. Designed for me. Outlining your perception of my predicament. And all in all, too precisely accurate for me to slip aside. I admit to a futile lament that I am not the embodiment of the Strategist and able to turn so swiftly off the dime. I am, unfortunately, real. The inventions you suggest I accomplish, I do not know how to invent. Though truly, I would. You struck so closely to the heart of me when speaking of family, I nearly found my failures unbearable. I near tears as I think of it. How is that possible? I am hardened to the world's slide into wholesale reactivity, and mine. Though I know I cannot shield my children forever, I can attempt no other thing; my sense of future has become the unclimbable wall you describe. Each attempt I have made to scale it has failed."

"So far." Melissa said calmly.

The Agent stared at her. His mind, so skilled at creating his mature self, had lost memory of the feeling of its own rapid reorganization and so instead categorized the quick changes in topology as confusion and weakness.

Melissa directed his attention. "If you will forgive, for a moment, the inadequate metaphor of the wall, you must see we are able to nullify it. Do you know what was on the other side?.

The Agent just waited.

"Nothing. Neither the wall, nor the terrifying future it pretends to hide, exists." Melissa paused to enable the Agent the opportunity to react.

"I wish what you mean was clear to me, but I am able to listen to an expansion of your thoughts and their application."

Melissa took up from her declaration that the future was not a predetermined ugliness waiting to deliver itself to the present. "Together these fictions proscribe imagination and act as a powerful debilitating inhibitor. Truly the wall is only a trick of mind; to scale it is to perceive through it."

"The wall itself is a product of imagination?" The Agent was always in search of inconsistency.

"It is a cumbersome paradox that fear imagined — aided by the hindbrain — can capsulate possibility imagined. So, for those entrapped by the wall — partitioned from possibility imagined — a futility only permits the present to push the wall ahead of itself, leaving just enough space every day for painting the actions of each moment. The future can be depicted only in our imaginations. When imagination is siphoned away by nameless fears, the true future leaks away to nothing. The wall is manifest in the mind by a fear of nothing. Then this nothing fear is used in various control schemes by manipulators of untrained minds."

"I am finding I react to the use of the word fear. It is potent and knowable at a certain gut level, yet, in the end, I am unable to describe its substance.

Again the Agent showed by example that he was patiently attending the content of Melissa's unpointed dialog. He himself was surprised by his behavior and attributed it to the oddness of present circumstance.

"Fear is an all-encompassing word. We mean to include a core synesthesia — attributes of the whole sensorium. We feel fear when we feel unsafe. The present has many ways to challenge safety. How does the removal of the wall enhance safety? It allows the future to be imagined. The most unsafe and therefore fearful behavior of mind is to limit possibility. Behavior that constrains choice, reduces safety and increases fear. By itself, misunderstanding is trivial. It requires only that the construct of understanding be adjusted to overcome deficiencies. But, when fear becomes formalized it is more difficult to deconstruct."

"Formalized?"

"...formalized, as when agencies adopt fear as a strategic or tactical tool, or when patterns of thought incorporate fear in secondary gain substructures. Religion and cynicism are just two examples."

"You have identified me as a cynic." The Agent was not disputing the correlation.

"No, only identifying you as a holder of the behavior of cynicism...this is a critical distinction. If we considered you to be defined as cynicism we would avoid dialog with you at this time. You would be a dictionary entry."

"I understand. I am not a definition." This was delivered without sensitivity to the fact that he was obviously in the throes of critical analysis by strangers. He had shifted so far away from the domain of effrontery and petty identity maintenance that he could listen freely for information of use, even about himself.

"It is cynicism that cements the illusion. Cynicism tries to substantiate the wall around fear and superstition, troweling it smooth with slick rationalization and biased evidence. One rationalization is the unfounded supposition that negative patterns are the primary element of nature, and therefore the future — an idea true only for those incapable of meta-analysis.

Melissa queried Thodkin via silent dialog. *"Am I over the edge? Can a cynic absorb such metaphors? Isn't it his nature to block consideration of the affliction?"* Thodkin gave a quick little nod to go on and in the same moment encouraged her to hold nothing back, *"Now is the time. You have wrested his attention from the grip of dispassion. He long ago transitioned from tolerant acquiescence to active listener. Admitting it or not, cynics are driven by emotion. They are driven by the conflicting knowledge that life is worth living, that few own that knowledge and that there is nothing those aware of the critical lack have been able to do to help. The path to change for a cynic is not through logic, it is through disconnecting them from their sense of failure and then reconnecting them to hope."*

She continued in what would seem to the Agent as an uninterrupted sequence of thought. "Cynics hone the surface with twitching strokes of fear animated by ominous shadows shifting in smoke-filled perceptions. They twitch in places where knowing what one is to become — if they were to shuck the calcified shell surrounding choice — is obscured. A mind taken over by the insidious force of such memes generates a seamless progression of the wall that inhibits open access to the possibilities of the future. But, the wall is a fiction, and this fiction can be dismantled, thereby eliminating one more foil of ignorance, superstition and entropy. Once removed, the hidden potentials of life open wide in nourishing, bright, color-filled panoramas before us. It is this dismantling that we have undertaken, in order to walk forward into scenery of neonature.

Melissa had challenged the Agent to keep pace with her — to stretch as she had, beyond the strictures they had absorbed from the institutions of learned behaviors. She was breathing quietly as she calmly connected to the Agent. She held his eyes. He sought for the untruth in her, the hyperbole. But he could not find it. He found only a deep pool of calm compassion, knowledge and intensity of focus. She was, he was sure, more than he had guessed. Somehow, she was the linchpin in everything the ToolShops had accomplished. She was not an ugly grub or liquid indebtedness to be distilled from a smarmy broth. She was honest and he could see she did not think of him as a failure. She only thought he was unfinished — an analysis he was able to consider only because it came from her.

After a quiet moment, Thodkin spoke again. "Can we agree that we have accomplished something you consider unthinkable?"

The Agent acknowledged: "I have admitted it."

"Then your assessment must be in error," Melissa observed.

"The wall is scalable or subject to dismantling," Thodkin clarified.

"It is," the Agent acknowledged again, only this time he was surprised to notice more than abstracted logic was supporting the assertion. The urge to add, "so it seems," was only a squeaky little gasping phrase rendered by a disintegrating behavior.

"So, good. Finally we can get to the point of this meeting. That is, besides a delightful lunch." Melissa was laughing as she spoke. "We called you here today to request that you join our team. We are about to enter a new phase in which you can be of tremendous value."

"This would be considered co-opting by some," the Agent observed.

"They would be incorrect. Only one definition of co-opting implies creating a false need to avoid complication or risk to an objective. You can do as you will. We will not interfere with you. Forgive my need for precision. If you must think of this as co-opting, it must be:

a : to choose or elect as a member or b : to appoint as a colleague or assistant." Melissa adjusted the conversation's environment to remove ambiguity.

"I find it difficult to imagine your needing my help. Can you clarify that?

"We can illustrate it this way," Thodkin answered. "Change most often takes place in an inverted pyramid, where the bottom point is the genesis of an idea or objective and all that is really present on the ground — a single point, a solitary root. There is really no such thing as a grass-roots movement. Those whose personalities and intellects are nearest to the ecology required to support an idea are the first to adopt it. These are numerically few, but they are the ones most able to create an adoptable ecology for those next in the progression. The next group is numerically larger and the cycle continues until it somehow reaches the numbers required to be classified as a grass-roots movement."

"Critical grass," observed the Agent with a little fleeting smile. "I represent some part of the growing pyramid."

"To use phrases that are largely meaningless for lack of specificity and support in sensory specific systems of evidence, you represent the entropic force of cynicism. Minds nearly as capable as yours, but caught in the annealed attractor orbit of a cynic model, are our next focus. We wish to develop courseware and infuse a global environment with it to enable a rapid, pervasive softening of the encrusted scab that is the cynicism model."

"You will tell the world of cynics the story of The Strategist?" The Agent was nearly laughing now. He was feeling a rare sense of lightness and vitality. "Please, I am sorry, it just struck me as almost possible. How is it that you felt qualified to approach me, if I am to be approached by someone similar but closer to the point of the idea? How is it that you know anything about cynicism? All evidence points to the contrary. No cynic would have begun to build a Rivendell II. It would never have occurred to them."

"It seems," said Melissa, "we have found an exception to the model. Though we will not threaten credulity by providing a complete explanation, I will offer you two thoughts. One, an obvious one that you already know to be true no matter how it is discounted, is that it is possible to learn from others' experiences. We can sometimes make the leap of understanding by virtue of relationships with other people sharing our world. We are, by far, more similar to others in the world than we are different. The second thought is…Thodkin was not telling you a story. He did not design it for you, but selected it for you."

"I have made a small study of cynicism and cynics. So, to the topic." Thodkin inhibited the Agent's time to digest Melissa's cryptic revelation. "I am soon to need a companion of skill, experience, intelligence and moral fiber to accompany various travels and assist operations across the planet. Melissa will no longer accompany me. We offer you the job."

Just so.

ASYLUM

In a Little Room an Inmate Expires

In a padded room of muted greens the body of a man sits in a corner, out of sight from the locked door's view port — it is a flaw in the room's design.

He was an intelligent man in the good times of his life. He garnered wealth, which he just confirmed was not possible to take with him. He had been sure he would need no wealth where he was going — and he really didn't care. With money, came noise and his single goal was silence.

The worst tragedy of his life was his lack of skill at adjusting to the tragedy of his life. Others had overcome such adversities as his, using many methods, sometimes the help of professionals, sometimes just a capable personality. But with energy this man chopped and chopped and after much circular tedious effort, felled his own tree before it had grown able to live beyond the fire of his personal Dante-esque canto. Now he lay very still; his body had begun to cool, one hand dropped beside him to the floor, his head nodded to a resting position and his life winked away. He did not attain the future this day. As for silence, he did not experience that either — that was left to others.

Or his life would have seemed to wink away, had anyone been observing him — monitoring his vital signs. For several minutes he sat motionless, heart stopped, mind lost from oxygen deprivation. Then just before irreparable damage to the brain cells was imminent, the heart began to beat again. The lungs restarted — their machinery of osmosis was reignited. And some very small devices began to appear within any capacious cavity. Soon the devices were distributed throughout the organism. Repairs were begun and the construction of a new mind was frenetically underway using available resources extracted from bequeathed neural networks.

It would seem to the fictional observer that the man had miraculously revived. Deduction would suggest that he had some immunity to the drugs he had hoarded all these months…drugs that would have put a horse to sleep for a last dream of open meadows and tall delicious grass.

Another Pleasant Green Room

In her own room again, after her session, the former president of Princeton relaxed. She felt much calmer these days and was noticing something of her former clarity returning as they reduced the dosage of the antidepressants. She knew that her faculties were impaired by the shroud the drugs placed over her feelings. Sooner or later she would need to live without them; maybe when she was able to clearly organize the events in the Lab as a cohesive memory. Maybe when she knew the green beast for what it really was. Maybe then she would request the stopping of the drugs. She had come a long distance from the blackness that had enveloped her that year so long ago. Someday she would entertain leaving the hospital. Someday for good. She could do it if she put herself to it.

The Crafts Room

His jailers, as he had internally characterized them, were pleased with themselves. After a terrible "bottoming out" requiring nearly four days in hospital care, their patient was making a remarkable physical and psychological recovery. It was the thing of bestsellers and scientific journals. After years of unpleasantness, leading to nearly total disintegration of his personality, he had suddenly begun to rebuild himself. There were a few unprecedented moments of delusion, like the early stages when he pounded the walls of his cell screaming that he was the author of the universe. But he had made it past the delusions and again knew who he was and what had happened to bring him here.

After his session, he was permitted to join others in the Crafts Room for the first time in his five years at the institution. He had learned that this was to be the first in a series of tests leading to supervised day trips and then a halfway house and finally to returning home. The man chuckled at himself. Next time he would not rely on the Quantum Analyst, a special intelligence he'd used to advise him regarding viable hosts. It was just too much the kidder. Ahh, well, there was nothing to do now but see it through — insulting the Analyst was to be avoided, as in not enjoying the joke would inspire a new, enlarged 'humor' from it.

The recovering man walked slowly about the cheerful room. It was much better than his green cell. He nodded politely to any who acknowledged him. Then, from a distance, he peered over the shoulder of a black woman making use of a small corner area lit with the natural light from the room's large windows. At his distance she seemed to have a talent and was painting a detailed scene in the minute brush strokes of a pointillist. It was painstaking work few artists could tolerate, let alone an inmate in this institution. He approached the woman, finally making a careful introduction, wishing not to create a stir.

"Permit me to interrupt you to express wonder at the work and image you are depicting. I know little of these things, but it seems to me that you have a marvelous talent."

The woman looked up at him. After years here, she could evaluate her fellow inmates instantly. This one was nearly fully recovered. He was not quite an open book, but he was not dangerous. She could interact with him as she would. "Thank you. I admit it is an image that haunts me...one key to my recovery."

The man looked at her with a surprising and unexpected quality of gentleness. In that moment he had learned something about nuance that all of his many years and adventures had not taught him. He understood the potentiated effects of his own actions in a Force Bottle III. Tears welled up in his eyes, streaming of their own accord down his cheeks and onto his pajamas. He only barely knew what tears were, though he realized he had heard of them.

He paused in a relentless concentration, fusing all his vast resources...inquiring of the Quantum Repository...scouring the Earth texts he had acquired for his library...applying any intelligences willing to allocate time in a research unprecedented on Earth or any universe known. He spent many fortunes he would have to someday earn for the single purpose of making the next sentence he uttered pay for the shame he deserved and to somehow give this woman the key she needed to complete her recovery.

She waited only a little longer than she expected for his completion of standard introduction protocol. She ignored the fountain of tears as the man finally spoke.

"It is a perfect replication of the synthesized organic, dubbed the Green Hair Comber. I take it you were unable to find

Melissa. As inadequate as it is, may I apologize for the incident and the suffering it has imposed on you?"

He was unsure of the success of his objectives as he caught the woman in her faint.

The End

Potential is not the same as achieving.
Pairing an objective with action is the same as achieving.

Achievement is another thing altogether.
And the hierarchy of values
—which is to say, deriving a rational schema of what is worth achieving and why—
is just a mess.
Still, delegation of the task is doom.

Doom is for sensation junkies.

— *Thodkin Marblemeister*

APPENDIX

Appendix A: Potato Salad

The Universe's Best Potato Salad…It's just that simple!:

3 Pounds: serves from 1 to 6 persons, depending…

Ingredients:

1 garlic cluster
3 pounds of Red Rose potatoes; select for uniform size about twice the size of an extra-large egg
1/3 cup red wine garlic vinegar
1/3 cup Mazola corn oil
1 table spoon salt
2 tablespoons freshly ground pepper
1 red onion (size of baseball)
8 baby gherkin dill pickles (baby sweet gerkins are also nice) (a stress on *baby*)
12 Eggs
3 tablespoons BEST FOODS MAYONNAISE (HELMANS MAYONNAISE if on the other side of the Rockies) Accept No Alternates, Do Not Produce False Images. Abandon project if required mayonnaise is unavailable!!!!

Cooking Utensils Required:

Obtain large salad bowl (preferably a yellow ceramic one for proper ambiance)
3-quart pot for each 3 pounds of potatoes
2-quart pot for eggs
Sharp knife, wide enough to smash garlic cloves (12-inch is good)
Tablespoon
One adult fork
Pepper grinder

Procedure:

Place potatoes—SKINS ON!!!— in pot, covering with cold water. Cover and cook for approximately 1 hour.

Place all 12 eggs in pot covering with cold water. Cover and cook for 20 minutes.

Note: Upon completion of egg cooking time, remove eggs from stove and place under cold water—immediately.

While Stuff Cooks

Slice one garlic clove in half.

Smear rigorously around every smidgen of the yellow salad bowl. Miss no area all the way to the rim.

Select 3 fresh garlic cloves

Place a clove under the FLAT part of your 12-inch knife (NOTE: The knife should be FLAT, never wallop the sharp edge of a knife) and wallop the thing. This will smash the clove. Remove the white skin and deposit the remainder into the yellow salad bowl.

Repeat the above step for cloves 2 and 3.

Add 1/3 cup of red wine garlic vinegar into yellow salad bowl
Add 1/3 cup of 1/3 cup Mazola corn oil into yellow salad bowl
Add 1 table spoon salt
Add 2 table spoons freshly ground pepper
Slice red onion into nickel-sized chunks
Add red onion pieces into yellow salad bowl
Slice 8 pinkie-sized baby gherkins (My pinkie size is 2.75 inches; 3 inches is Okay. The main thing here is that the pulpy internals of large pickles are to be avoided. Use only nice firm nubile pickles if substitution is required.)
Add sliced gherkins to the mess in the yellow salad bowl
Mix the ingredients currently in the yellow salad bowl

You are doing well so far...success is not yet imminent, but you should enjoy your achievements to date.

Upon completion of all the slicing and such, peel the eggs. Half of the eggs may be used for deviled eggs. I recommend this...it is always a nice way to honor your guests. If you have good friends coming over to dine...cook more eggs and devil quite a batch. If you just have acquaintances coming over...save the deviled eggs for later personal consumption, or eat them before the guests arrive. See a simply delightful deviled egg recipe below as a bonus.

It is my experience that following the order I have outlined requires about 48 to 54 minutes. This means that you can now check the potatoes. A properly cooked potato will show splitting skins. Take an adult fork and gently slide it into one of the larger potatoes. If the texture is firm and smooth—not requiring much force—the potatoes are ready for removal and peeling; IF NOT SMOOTH ALL THE WAY TO THE BONE—ALLOW CONTINUED COOKING...There is nothing worse than crunchy undercooked potato salad, like you get in crummy delicatessens. If that is what you wish to achieve, why bother; stop right now and go buy some of the tasteless crunchy glop. You may also give this book away and seek a family more to your liking...say the Bumsteads.

The peeling is trickier than you might think. It is important to fold hot or at least very warm potatoes into the yellow salad bowl. So I offer you a means for peeling fresh-off-the-stove potatoes.

After removing the potatoes from the fire, pour out the hot water. Remain clear of the steam...it can provide a second degree burn before you feel it.

Next, fill the pot...potatoes still resident, with cold water.

Take one potato at a time and, holding the potato with an adult fork held by your non-dominant hand, scrape the skin off of the potato with the tablespoon held in your dominant hand.

Lastly, cut the now peeled potato in eighths and place into the yellow bowl.

Peeling must be accomplished steadily and quickly to prevent the first potatoes from sucking up all the marinade. To avoid this, if you are a slow peeler, place the peeled potatoes in a temporary holding bowl until all the potatoes are peeled…then place the entire set of peeled potatoes into the yellow bowl.

Boy, things are really getting exciting now. And you are really doing terrific. The next few steps will be your crowning acts.

Now that you have all the potatoes in the yellow bowl, gently fold the potatoes into the previously resident ingredients. Please be gentle…we can make mashed potatoes some other time; right now we are hoping to have some chunks of potato still recognizable as such when all the folding is completed.

When the potatoes have a nice sheen to them (they are evenly integrated with the other good stuff). Slice at least six of the eggs into eighths (two lengthwise cuts and one in the width) and fold in the eggs.

If pride were money, I'd be a millionaire. I am so proud of all that you have accomplished so far…really…you are doing great. Now, however, you must make a decision. Do you have time or are you going to serve the delicious meal warm? If you have no time, you will want to complete the following steps. If you have 3 or 4 hours before serving, you will want to wait 2.5 or 3.5 hours, or 1/2 hour before serving to complete the next steps. You choose; you have come this far exhibiting a special culinary altruism…I believe in you.

Final Steps:

Best Foods Here We Come! Caution: Mayonnaise is sneaky. You will probably find that 3 heaping tablespoons full of Best Foods Mayonnaise is just right… but at first it will seem like too little. I leave it to you, but consider yourself forewarned. The difference between a greasy, sloppy disappointment and a moist richly flavored delight is a tablespoon of mayonnaise.

Fold in the 3 heaping tablespoons full of Best Foods Mayonnaise…gently. Try not to break a single potato…that's the art of it.

The final tune-up:

Take a bite.

Add a little pepper to get a proper snap.

Add a little salt to sharpen the flavor identities

Repeat the three previous steps in the tune-up until it is just right.

NOTE: If you have no taste-buds, leave this to your guests. It's already the best potato salad in contemporary society…you'll get raves even without a tune-up.

Eat hearty, you deserve more than one helping for all that you have accomplished. You are the best!

Post Script:

Many people claim that which is properly the right of another to claim. Here there is no attempt at such usurpation. The above recipe was taught to a son by a mother…mostly. She was taught it by her mother and so on, until history meets with the nameless archetype German Potato Salad Inventor. The original was surely invented during a potato famine — that is, a time when there is little to eat but potatoes. This proves the common knowledge that out of adversity good things can come.

The above recipe, minus the precise technical development instructions, results from a woman, we'll call her Dolores, visiting a son in a foreign land. She presented him a gift of a favorite meal. He chose potato salad. Not just any potato salad, the potato salad he dreamed of, longed for, and felt lost without. During this prototype fabrication, the young man sat by, pencil in hand, making careful notes. Each grain of salt was measured. If she added, he added, if she tuned, he tuned. If she did a little incantation, he recorded it. As a result, he now consistently makes a great potato salad. It is so great that restaurants have refused the challenge of a customer taste test 3 summers in a row. One can feel sure this is because of a deep-seated fear of the omnipotence of the above recipe and construction methodology.

Others claim to make excellent potato salad. No breath will be wasted to dispute them. Let the unmitigated phantasmagoria experienced by flask-shaped bodies — chiefly in the epithelium of the tongue, which are the end organs for the sense of taste — when touched by the above concoction resolve the disputation. You have but to assiduously follow a few steps to learn a valuable and unassailable truth. These are rare in our time…wouldn't you enjoy one?

P.S. A very simple Devil is a better Devil:

Sometimes the purpose of a good thing is lost in the flourishes that surround it. So it is with deviled eggs. Purists know that deviled eggs should taste predominantly of egg. To return to the roots of your culture, try this:

Slice 6 or more hardboilded eggs in half longitudinally (the long way, with attitude additionally…longitudinally). Segregate the white shells from the yolks—yolks in a bowl, whites on a nice serving dish. If you have one of the dishes with egg dents in it, so much the better.

Take 1 tablespoon of BEST FOODS MAYONNAISE (If not available at the time of this reading, consult archives for recipe and make it yourself) and mix it in with the egg yolks using a nice mashing adult fork.

Add 1 teaspoon salt

Add 1 tablespoon pepper
Add 5 shakes of Worcestershire sauce.

Tune-up:

Take a small bite.

Add a little salt to sharpen the egg flavor identity

Repeat the 2 previous steps in the tune-up until it is just right.

Scoop a generous amount of the yellow into the egg white halves.
Sprinkle paprika lightly over the finished deviled eggs. The paprika provides the devil's color and cannot therefore be omitted.
Sample as many as required to satisfy that longing for the good ole days.

Well, that's it...serving and eating are the next logical steps.

PEACE

ABOUT THE AUTHOR

(Image may appear younger than reality)

CI_132 L0ARII.

Scott Ellison has been building things for years. Often wandering beyond the leading edge, he has built factory automation systems, computer games, toys, software products, electromechanical widgets. He has wandered other activities too, like flying hang-gliders and helicopters. He can still knock the stuffing out of a tennis ball (not too difficult really, they have no stuffing). And, for fun, he has written many Thodkin Marblemeister extremely short stories. Thodkin's Spear is his first Thodkin book.

As a professional technologist, Ellison has predicted the march of many advancements and now sees formidable technologies accelerating past human ability to cope with them: if the advance of science will not be slowed (and it really can't be), something else must speed up. That "something else" is humankind's ability to reason about complex issues. In his first Force Bottle Chronicle, Ellison pulls no punches: the human limiting factor in our ability to reason is personality. In Thodkin's Spear, Ellison skillfully deploys metaphor, allegory, science and storytelling to make a case for the key role of personality in meeting humanity's challenges of the near future. It is a hopeful and optimistic voyage built upon a deep understanding of both technology patterns and human nature.

Made in United States
Troutdale, OR
11/03/2026

56844423R00159